HISTORIC HIGHWAY BRIDGES OF OREGON

HISTORIC HIGHWAY BRIDGES OF OREGON

Dwight A. Smith   James B. Norman   Pieter T. Dykman

OREGON HISTORICAL SOCIETY PRESS

Cover: Shepperds Dell Bridge, Columbia Scenic Highway, 1915. *(OHS neg. no. 65330)*

Frontis: Yaquina Bay Bridge, Newport, Oregon. *(OHS neg. no. OrHi 76880)*

```
Ø/NW 388.132 1989 c.1
 Smith, Dwight A.    20 00
 Historic highway bridges of
  Oregon
```

Second edition, revised, 1989

Originally published by the Oregon Department of Transportation, 1986.
This volume was produced by the Oregon Historical Society Press.

Copyright ©1989, Oregon Historical Society
1230 SW Park Avenue, Portland, Oregon 97205

All rights reserved. No part of this publication may be reproduced or transmitted in any form or by any means, electronic or mechanical, including photocopying, recording, or any information storage or retrieval system, without the permission in writing from the publisher.

Printed in the United States of America.

ISBN 0-87595-205-4

This book is dedicated to the bridge designers and builders--past, present, and future--for their imprint on the Oregon landscape.

CONTENTS

	Page
PREFACE	9
INTRODUCTION	11
HISTORIC PERSPECTIVE	17
Development of Bridge Technology	19
Bridge Building Tradition in Oregon	27
INVENTORY AND EVALUATION	37
Need	39
Purpose	40
Methodology	41
Evaluation and Identification	42
Coordination and Determination of Eligibility	47
HISTORIC HIGHWAY BRIDGES	51
Study Identified Bridges	56
Truss Bridges	57
Arch Bridges	86
Suspension Bridges	112
Moveable Bridges	114
Slab, Beam, and Girder Bridges	121
Old Columbia River Highway Bridges	132
Covered Bridges	161
Other Historic Bridges	206
CONCLUSION	217
BIBLIOGRAPHY	223
GLOSSARY	229
APPENDICES	237
Appendix A: Guide to the Bridge Photo-Description Pages and State Highway Numbers	239
Appendix B: Significant Bridge Designers and Contractors	242
Appendix C: Bridge Statistics and Location by County	248
Appendix D: Structural Type and Number	256
Appendix E: Reserve Bridges	266
Appendix F: Notable Highway Bridges Constructed After 1940	296
Appendix G: Master List of the Inventoried Bridges	301
INDEX	313
PHOTOGRAPHY CREDITS	323

FIGURES

Number	Title	Page
1	Shepperd's Dell Bridge (1914), Old Columbia River Highway	14
2	Basic Bridge Types	20
3	Common Historic Truss Designs	22
4	Truss Connections	23
5	Willamette River Suspension Bridge (1888), Oregon City	24
6	John Day River Bridge (1918), Astoria	25
7	Rogue River Bridge (Ca. 1890), Grants Pass	26
8	Toll Bridge Rates	27
9	Camas Creek Bridge (Ca. 1900), Ukiah	28
10	McKenzie River (Booth-Kelly) Railroad Bridge (1882), Springfield	29
11	Willamette River (Morrison Street) Bridge (1905), Portland	30
12	The Bridges of Portland (Aerial View)	31
13	Coos Bay (McCullough Memorial) Bridge (1936), North Bend	32
14	Columbia River (Interstate 205) Bridge (1982), Portland Area	35
15	Evaluation Criteria	40
16	Rarity and Uniqueness Criteria (Details)	44
17	Chasm Bridge (1937), Tillamook County	45
18	Artistic Embellishments (Details)	46
19	Rhea Creek Bridge (1909), Morrow County	48
20	Historic Highway Bridges List	53
21	Yaquina Bay Bridge (1936), Newport (Drawing)	56
22	Crown Point Vista House (1918), Columbia River Highway	132
23	Standard Covered Bridge Plan (1934)	161
24	Coquille River Bridge (1922), Coquille	220
25	Grande Ronde River (Old Rhinehart) Bridge (1922), Union County	221

PREFACE

It is with great pleasure that ***Historic Highway Bridges of Oregon*** is published. The Oregon Department of Transportation (ODOT) hopes that it appeals to a wide audience. The document is, foremost, the result of a department research study and will assist in transportation planning and development work. Additionally, the book is a source document for future bridge research, being the first comprehensive inventory of one of Oregon's engineering features. Finally, the general readership is provided a detailed view of the historic bridges of Oregon. Over two hundred photographs and descriptions illustrate the rich legacy of bridge building in this state.

The historic bridges study involved the time and talents of many department staff and the support and assistance of private citizens and other agencies. Some years ago, Lewis L. McArthur, a respected authority on historic engineering structures, suggested this study to the Department of Transportation. Lew's concern that historically important engineering features were being replaced was shared by the department. The proposal to inventory the historic bridges in Oregon was a timely complement to the need to address department compliance with historic preservation laws. Consequently, a committee of ODOT Environmental Section and State Historic Preservation Office staff developed a study framework and secured the support of the department's administration and funding assistance from the Federal Highway Administration. The Environmental Section was given the responsibility for implementing the study.

The principal author of ***Historic Highway Bridges of Oregon***, Dwight A. Smith, cultural resource specialist in the Environmental Section, was the study's manager and diligent workhorse. Dwight's unfailing enthusiasm for historic bridges, along with his creativity and professionalism, carried the study through many years. James Norman, also of the Environmental Section, served as the research associate for the study, was the principal photographer and coordinated the design of the book. Pieter Dykman, the environmental research supervisor, had the responsibility and pleasure of supporting the effort of these two talented individuals and reviewing the results.

The study was conducted with the early support of the State Historic Preservation Office and their continuing assistance through the review of the study results. David W. Powers III and Elisabeth Walton Potter offered valuable advice and professional expertise.

Three private citizens and bridge experts donated considerable time and knowledge to the evaluation of the historic bridges: Lewis L. McArthur, Vice President, Ray Becker Company, Portland; Louis F. Pierce, President and Chief Engineer, OBEC Consulting Engineers, Eugene; and Thomas J. McClellan, Professor Emeritus, Oregon State University Engineering Department, now with CH2M-Hill, Corvallis. Their knowledge and appreciation of historic bridges added depth and substance to the bridge evaluation.

Many members of the Department of Transportation assisted the study. Walter J. Hart, State Bridge Engineer, aided the research on many occasions and served on the evaluation team, along with Jack L. Davis, also of the Bridge Section. Gerald W. Test assisted in the early exploratory stages of the study. Jerry Robertson, Craig Markham, and others assisted with field inspections and photography. The excellent design and artwork of this publication were contributed by John Davenport, George Kraus, and others in the Graphics Unit. The State Printing Division printed the document.

Like family, the final credit and greatest praise for support goes to our fellow Environmental Section staff. Probably no one escaped giving some aid, advice or solace in the course of the study. Many persons assisted in various reviews. Lori Butler, with other clerical staff assistance, typed all of the innumerable documents produced during the study speedily and accurately. Campbell M. Gilmour as section manager provided continuing encouragement and administrative support to the study, as did his predecessor, Gary A. Potter.

To all of these outstanding individuals mentioned and others inadvertently omitted, we extend our most sincere thanks and appreciation.

Pieter T. Dykman
Environmental Research Supervisor
Oregon Department of Transportation
Salem 1985

INTRODUCTION

North Fessenden Street Overcrossing (1909), Portland, Multnomah County

INTRODUCTION

One does not need to be an engineer to appreciate the idea of a bridge, or its beauty; there is no more overt, powerful, or rational expression of accomplishment--of man's ability to build. Bridges are among the finest examples of American structural art--powerful objects of pure utility and science--a realm in which Americans have excelled.
 David Plowden, **Bridges: The Spans of North America**

Bridges are among the most ancient and honorable members of society, with a background rich in tradition and culture. For countless generations they have borne the burdens of the world, and many of them have been great works of art.
 Charles S. Whitney, **Bridges: Their Art, Science, and Evolution**

In recent years there has been increased interest in old bridges as historic resources worthy of recognition and preservation. While covered bridges have enjoyed attention for many years, the new interest in old bridges encompasses all types of structures important in the history and heritage of bridge engineering. Early metal trusses, arches, moveable spans, suspension bridges, and slab, beam, and girder structures are now sharing the awareness and appreciation of the public.

Old bridges are important parts of our cultural heritage. As historic legacies they are as important as the early pioneer cabin, the Victorian courthouse, and the string of "commercial palace" buildings on Main Street.

Among the highway travelers' most vivid memories are those formed by bridges. Bridges create a sense of passage, open up wide vistas, and frame views. The appearance of an old bridge can suggest a sense of time and place and a different life style. Bridges may also appear as monumental landmarks, the identifying image of a city or town, or as simple, understated and elegant testimonials to good design and engineering. These designs of the past often reveal an exceptional sensitivity to scale and to their surroundings (Figure 1). Today, they are also rich in educational and symbolic meaning.

Oregon contains about 1,200 highway bridges in city, county, and state ownerships which were built before World War II. These early bridges are of various types and materials, ranging from simple timber spans across small streams to multi-span steel and concrete structures over major rivers and estuaries. These bridges were built in the early days of the automobile era, when vehicles were light and not very numerous. Today, many of these bridges must carry loads far in excess of their original design specifications, including heavy trucks, large numbers of automobiles, and farming equipment. Some bridges cannot meet these demands and are considered unsafe or obsolete.

While recent national and state emphasis has been on safety and the general improvement and replacement of bridges on our highways, historians and other members of the public are increasingly concerned that historically significant bridges are rapidly being lost. The issue, safety versus preservation, creates difficult bridge planning problems. This historic highway bridges study was initiated to assist in the resolution of this conflict.

This document presents the results of the historic highway bridges study conducted by the Oregon Department of Transportation. The study document contains four major sections--historic per-

FIGURE 1. *The Shepperd's Dell Bridge was constructed on the scenic Columbia River Highway in 1914 and was designed by the newly-created State Highway Department (1913). The structures on the Columbia River Highway established a tradition for excellence in design and compatibility with the environment.*

spective, inventory and evaluation, historic highway bridges, and conclusion. The historic perspective section provides the context for identifying important bridges and includes a discussion of the general development of bridge technology and the bridge-building tradition in Oregon. The inventory and evaluation section focuses on the study need, purpose, and methodology, including the evaluation procedure for identifying the historically significant bridges. The historic highway bridges section illustrates the 145 bridges in Oregon which are currently eligible for or listed on the National Register of Historic Places. The bridges are presented in four groups, based on the actions which resulted in the bridges being declared historic: study identified bridges, old Columbia River Highway bridges, covered bridges, and other historic bridges.

A bibliography and a glossary follow the basic report. Several appendices contain information on bridge designers and contractors, the locations and types of the bridges, photography and descriptions of other bridges of historic interest, a list of all bridges inventoried for the study, and other information. At the end of the document is an index and photography credits.

HISTORIC PERSPECTIVE

Moffett Creek Bridge (1915), old Columbia River Highway near Bonneville, Multnomah County

HISTORIC PERSPECTIVE

Bridge building is a rich and fascinating chapter in the history of technology and engineering. In Oregon, bridge construction also contributed to the general growth and development of the state. An examination of the general history of bridge design and construction and that tradition in Oregon provides a context for the understanding and appreciation of historic bridges.

DEVELOPMENT OF BRIDGE TECHNOLOGY

The history and development of bridge technology is a blend of the adaptation of new materials and improved designs to the three basic types of bridges: arch, beam, and suspension (Figure 2). The concepts behind the basic bridge types have been understood and used for centuries, but until recently the limitations of the available materials severely restricted length, capacity, and design.

There are two major kinds of stress in a bridge: compression, which pushes on or shortens a structural member; and tension, which pulls apart or lengthens a member. The arch acts in compression, the suspension bridge in tension, and the beam or truss (an open-frame beam) in a combination of tension and compression, to support the weight of the bridge and the applied load. The physical nature of bridge materials determines their ability to withstand these stresses. All common bridge-building materials, stone, timber, iron, steel, and concrete, are good in compression. The development of materials that are good in compression and tension, particularly wrought iron and steel, led to rapid advances in bridge design and technology in the mid-19th century.

Early bridge builders generally used whatever material was readily available. The first bridges of primitive man were probably crude stone slabs or logs laid over narrow chasms. In Asia and South America, suspension bridges made of vines or ropes of relatively long length were used. The nature of the material, as well as the current fashion and style, determined the design. Stone, for instance, was found to be naturally strong in compression, but with little or no strength in tension, and consequently made a poor beam structure. When properly crafted, however, it allowed for the construction of exceptionally strong and durable arches. Stone also proved an excellent material for foundations and piers. However, the many deficiencies of masonry construction, such as the massive expenditures of time and effort, the scarcity of appropriate stone, the unsuitability of many sites for short-span arches, and the need for skilled craftsmen, eventually forced man to seek more economical construction materials.

Timber, though obviously not as strong or durable as stone, has both compressive and tensile strength. This advantage, along with its almost universal availability, made wood a good choice as a bridge construction material. The length and quality of available timbers, however, limited individual span lengths in early timber bridges. Though the most basic of design concepts, the timber beam still is in use today.

Arch Bridges

The invention of the arch was a significant step in the history of technology. How and when the arch was discovered remain a matter of conjecture, but both the Sumerians and Egyptians had arched windows and vaults by about 3000 B.C. These early arches, constructed of wedge-shaped stones or bricks, were a great improvement over the earlier-used corbel (or false arch), because arches allowed wider openings and stronger support for the area above the openings. (With a corbel, each course of stones or bricks on either side of an opening is laid projecting slightly further than the one below, until the two sides meet at the crown.) Many centuries passed, however, before the arch was applied on a large scale to the construction of bridges, awaiting the genius of the Roman engineers.

The Romans were the first great bridge builders. The Romans were ambitious builders and constructed thousands of bridges, usually simple timber ones. It was their large stone arch bridges, however, that have endured as testimony to their craftsmanship. The durability of the early Roman bridges is shown by the survival of several 2,000-year-old stone arch bridges in Italy, Spain, and Portugal. The development of the arch bridge by the Romans may be considered the first adaptation of scientific principles to bridge construction.

FIGURE 2
Basic Bridge Types

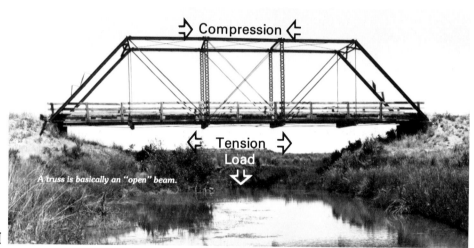

BEAM

SUSPENSION

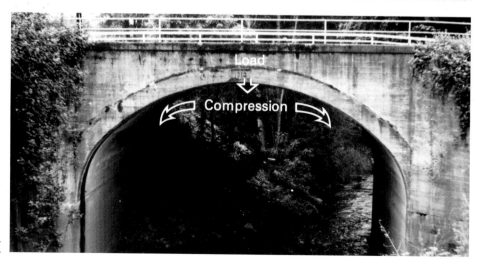

ARCH

After the decline of the Roman Empire, few bridges of importance were built in the Western world until the 12th century. Many notable arch bridges were built by religious orders and became sites of shelters and hospices for travelers. Bridges at strategic locations were frequently fortified for defense. The original London Bridge (1209) was lined on both sides by overhanging buildings and also included a chapel.

The Renaissance period transformed the arch bridge. Prominent structures became expressions of the art and architecture of the period. Architects, coming into their own as an artisan class, developed a number of styles--classical, rococo, baroque--and applied these styles to bridges, as well as buildings. Bridges became adorned with detailing, such as open balustrades instead of solid parapets, niches for statues within piers and abutments, chamfered edges on the voussoir stones to emphasize the pattern, and decorative panels and classical columns on the piers and parapets. Some of the most famous historic bridges are arches from the Renaissance period, including the Ponte di Rialto (1591), Venice, and Pont Neuf (1604), Paris. Masonry and masonry-faced arch bridges and aqueducts remained the recognized engineering approach to bridge building until the 18th century.

Truss Bridges

Modern bridge technology began with the advent of the truss bridge. Trusses were commonly used in cathedral construction in the medieval period, and their adaptation to bridges was inevitable. Andrea Palladio, an Italian architect in the 16th century, is credited with first adapting the truss for bridge building and provided truss bridge designs in his *Four Books of Architecture*(1520). Using the triangle, he designed three forms of trusses built entirely of timber. Swiss and German builders later adapted Palladio's idea or independently invented the truss bridge.

The truss bridge was developed and refined in America in the late 1700s and early 1800s. The first trusses were timber and covered to protect the bridge members from the elements. Famous pioneer bridge builders during this era were Timothy Palmer, Theodore Burr, William Howe, Stephen Long, and Ithiel Town.

Cast iron was used in the construction of arch bridges as early as 1775, but its brittleness and low tensile strength restricted it to compression applications. No significant changes occurred until wrought iron was introduced as a bridge construction material at the beginning of the 19th century. The production of high-grade wrought iron allowed the design and construction of all-metal trusses and composite truss bridges. Composite truss bridges used timber beams for the compression members and iron rods for the tension members.

Several truss designs of the mid-19th century became important in bridge technology, and many continue to be used today (Figure 3). Three of the most notable designs are the trusses patented by Thomas and Caleb Pratt, William Howe and James Warren. The Pratt truss, which placed the vertical members in compression and the diagonal members in tension, has been used in thousands of bridges across America. The Pratt design was best adapted to all-metal construction and spawned a collection of variations, the most important being the Whipple truss (a double-intersection Pratt) and the Parker truss (a Pratt truss with polygonal or curved upper chords).

The Howe truss, which placed the vertical members in tension and the diagonal members in compression, was well adapted for use with timber and was widely used to construct covered bridges. The Warren truss, designed by British engineer James Warren, put both compressive and tensile stresses in the diagonal members, simplifying the configuration and allowing for the fewest structural members. This design, first used in 1848, facilitated stress calculation and quickly became the most common of all truss configurations. Virtually all trusses manufactured today are variations of the Warren design.

The first metal truss bridges were constructed of wrought iron, but high grade steel eventually replaced iron as a building material. (Steel is stronger than wrought iron but nearly indistinguishable in appearance.) Prior to 1885, bridge builders relied on wrought iron. Between 1885 and 1895, a boom in the United States steel industry led to the construction of a mixture of wrought iron and steel structures. By the turn of the century, steel

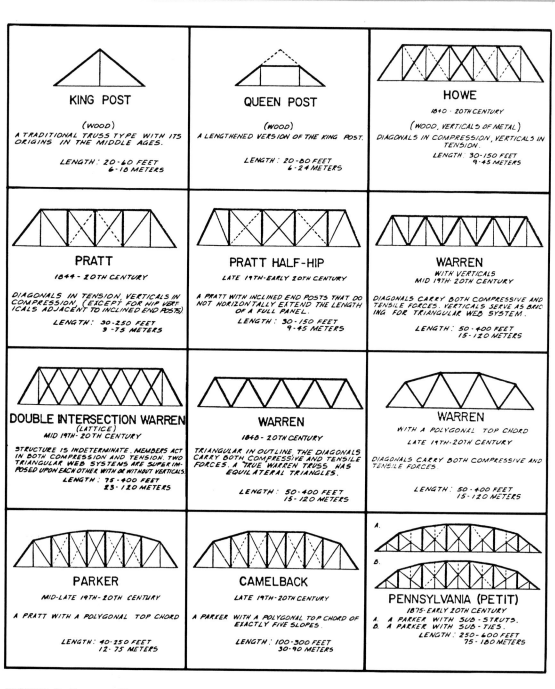

FIGURE 3. *Common Historic Truss Designs in Oregon.* (Source: Historic American Engineering Record, National Park Service, United States Department of the Interior, Washington, D.C.)

had replaced wrought iron for bridge building.

The truss members of the early all-metal bridges were connected with steel pins or bolts (Figure 4). These pin-connected bridges had the two-fold advantage of easy on-site construction and reduction of stress at the joints because the pins were free to rotate. Pin connections also facilitated the relocation of bridges, a common situation for many trusses, because of the ease of disassembly and assembly. However, the flexibility of the joints meant swift wear of the pins and truss members. The rigid riveted joint was developed at the beginning of the 20th century and increased both the load capacity and the service life of metal bridges. By about 1915, the riveted joint completely displaced pin connections.

Suspension Bridges

The suspension bridge (Figure 5) was evolving at the same time as the development of the truss in the 19th century. Iron chain suspension bridges were used in the Orient as early as the 7th century and in Europe by the mid-18th century. In the early forms, the roadway decking was laid directly on the cables or chains with little or no stabilization. In 1801, an American, James Finley, constructed the first suspension bridge to suspend a level roadway beneath the chains and later patented a bridge of wire cables.

French, American, and Swiss engineers were quick to utilize Finley's ideas. The Menai Straits Bridge, constructed in 1826 in North Wales by

Riveted gusset plates form rigid truss joints which decrease the wear on individual bridge members and extend the service life. The riveted joint made the pin-connected joint obsolete.

Pin connections allow easy on-site truss assembly and are generally found in Oregon only on bridges constructed prior to 1915.

FIGURE 4. Truss Connections.

Thomas Telford, expressed the importance of the suspension concept with an unprecedented 580-foot span. In 1834, a suspension bridge with a clear span of 810 feet was erected at Fribourg, Switzerland. The suspension bridge was established as the dominant choice in situations where long spans were required. By comparison, arch bridges of the same era rarely exceeded 200 feet in span, while trusses were capable of a span of about 400 feet.

Charles Ellet designed and constructed a suspension bridge across the Ohio River in 1848 with a span of 1,010 feet. The 1,595-foot Brooklyn Bridge, a monumental engineering feat, was constructed between 1867 and 1883 by John A. Roebling and his son Washington. Since that time, advances in cable technology have allowed suspension bridges to span unprecedented lengths. The world's longest bridge span is currently the main span of the Humber Estuary Bridge, between Hessle and Barton, in England, at 4,626 feet. Opened in 1980, the suspension bridge has twin towers, 533 feet tall.

Moveable Bridges

Moveable bridges were developed in response to the need for providing high or unlimited vertical clearance for river navigation. The three general types of moveable span bridges are the swing bridge, balanced on a central pier and rotating about a vertical axis; the vertical lift, which raises vertically in a horizontal position; and the bascule, developed from the medieval drawbridge, which swings upward (Figure 6).

Moveable bridges underwent rapid change as technology advanced in the 19th century. When the railroads and canal systems spread across the United States in the 1830s, the demand for moveable span bridges increased correspondingly. The early moveable bridges were all swing designs. Five bridges built across the Charles River in Boston in the 1830-40 period were among the earliest swing spans and consisted of crude timber trusses hinged at one corner. The center pivot swing design was developed about 1850 and was used almost exclusively until the end of the century.

FIGURE 5. A suspension bridge across the Willamette River was built in 1888 at Oregon City. One of only a few large suspension bridges in Oregon's bridge-building history, the narrow bridge was replaced in 1922 by a steel arch span.

FIGURE 6. *An early small bascule bridge in Oregon was the John Day River Bridge, constructed on the Lower Columbia River Highway near Astoria, Clatsop County, in 1918. The timber truss bascule opening span was flanked by two timber truss covered spans.*

Around 1890, practical methods of counterbalancing the enormous weight of the span and refinements to the electric motor led to the development of the modern lift and bascule bridges. When the bascule and lift spans were introduced, the swing span tended to disappear because the new moveable types opened faster. The first major lift-span bridge in the United States, the South Halstead Street Bridge in Chicago, was built in 1892. It had a 130-foot moveable span which could be raised 155 feet. The first bascule appeared in its modern form in Chicago in the following year on the Van Buren Street Bridge.

Reinforced Concrete and Modern Bridges

While engineers were refining and perfecting the various forms of arch, truss, suspension, and moveable bridges in the 19th century, the development of steel and reinforced concrete set the stage for another revolution in bridge design. Concrete had been widely used in piers and foundations, and steel soon dominated truss construction because of its increased strength. It was, however, the blend of steel and concrete that allowed bridge construction to add new dimensions and designs. The plasticity inherent in concrete was combined with the strength of steel to form a structure of virtually any configuration.

Reinforced concrete was initially used to refine the arch bridge. The first use of reinforced concrete for bridge construction in the United States was the Alvord Lake Bridge (1889), a small arch structure in Golden Gate Park, San Francisco, still in use today. This new material allowed for extremely flat arch configurations and greatly increased span lengths. Bridge designers quickly responded to the capabilities of reinforced concrete and produced imaginative and graceful designs able to bridge virtually any site. These expressive combinations of engineering and art are seen in the work of Robert Maillart (Switzerland) and Eugene Freyssinet

FIGURE 7. *Typical of the early metal high truss bridges in Oregon is this span across the Rogue River at Grants Pass, circa 1890. (Photograph courtesy of the Josephine County Historical Society.)*

(France). In Oregon, state bridge engineer Conde B. McCullough designed outstanding reinforced concrete arch spans.

The reinforced concrete arch designs were modified as construction and labor costs rose and artistic styles changed during the 20th century. Ornamentation on the structures slowly disappeared, and bridges with clean, simple lines appeared. Economy of materials became a primary concern, particularly during World War II. The new bridges created an aesthetic standard of their own, expressed in the overall concept and flow of a design and its harmony with the environment.

In addition to the arch, reinforced concrete was also introduced as a building material at the beginning of the 20th century for slab, beam, and girder bridges. Steel and concrete slab, beam, and girder bridge designs have become the predominant choice for most contemporary structures, including the new generations of precast concrete box girder bridges.

BRIDGE TOLLS

For every wagon and a single yoke or oxen or span of horses, mules, etc.	25 cents
For a single horse, and carriage	18 cents
For a single footman	5 cents
For a man and horse	10 cents
For all other horses and cattle, per head	3 cents

FIGURE 8. Toll bridge rates fixed by the Oregon Provisional Government in December 1845. (Source: Lee H. Nelson, *Oregon Covered Bridges*, Portland: Oregon Historical Society, 1976.)

BRIDGE BUILDING TRADITION IN OREGON

For the most part, bridge building in Oregon followed the trends and developments occurring in bridge technology across the nation. However, Oregon's network of large rivers, estuaries, and creeks provided the need and opportunity for a large number of bridges, and a rich bridge-building tradition has resulted (Figure 7).

Pioneer Settlement and Early Development, 1840-1870

From the earliest times, the people of Oregon, Indian and pioneer alike, used water transportation whenever possible. When the Lewis and Clark Expedition arrived in 1805, the Columbia River was already an important transportation route with hundreds of dugout canoes, some splendidly carved, plying the river.

Until 1841, overland migration to Oregon was negligible, being mostly adventurers, explorers and hunters, a few of whom remained and settled, and a small number of missionaries. The major settlement awaited the opening of Oregon's premier "highway"--the Oregon Trail. From 1841 to 1869 approximately 350,000 people immigrated to Oregon on the Oregon Trail. Caravans of covered wagons made the 1800-mile trek across the western half of the continent from Independence, Missouri, to Oregon City in the Willamette Valley, in about five months. Major rivers and streams were forded or crossed using small crude boats. At The Dalles in eastern Oregon the hardy immigrants came to the end of the overland trail. The imposing gorge of the Columbia River was impassable for wagons. Livestock had to be driven over the Cascade Range, the wagons dismantled and the last 100-mile leg of the journey made by raft or Hudson's Bay Company bateaux on the Columbia River to Oregon City and the Willamette Valley. The Barlow Trail, a land alternative to the Columbia River route, was blazed over the Cascades south of Mount Hood in 1845.

Bridge construction was not initially a major interest in the Oregon Country. The settlers were more concerned with survival, building cabins and fences, and breaking ground. Water transportation on the Willamette River, the Columbia River, and other streams was convenient. Canoes and rowboats were the first vessels. Keelboats were introduced in 1846, and steamships began regular service in the 1850s.

Ferries were essential to transportation in early Oregon before bridges spanned the major waterways. These ferries were usually operated as a private enterprise and many continued in operation well into the 20th century. The earliest recorded ferry operation was built by Jesse Applegate in 1844 across the Willamette River, north of Salem at the Willamette Mission. Switzer's ferry was established in 1846 across the Columbia River from Fort Vancouver, and there were ferries across the Willamette River at Portland as early as 1848. In addition, a ferry that figured prominently in Oregon history was Olds' ferry across the Snake River, established in 1862. (The existence of Ferry streets in many Oregon towns and cities attests to the large number of early ferries.) Today only three ferries remain in service in Oregon. These are all county-owned and ply the Willamette River at Buena Vista, Wheatland and Canby. A Washington ferry still crosses the Columbia River between Cathlamet, Washington, and Westport, Oregon.

Land transportation in the 1840s and 1850s was limited largely to horseback and ox-drawn wagon. Short roads only led from the farms to the nearest river landing. As the population of Oregon increased in the late 1840s, the need for roads and bridges became apparent. One of the early actions of the Provisional Government (1843-48) was to appoint a commission to establish roads. A number of routes were opened and improved through public subscription. The government also authorized the incorporation of several toll bridge companies (Figure 8). A few years later after Oregon became a territory, the provisions of an 1849 act placed bridge construction under the aegis of county governments. Each county court was empowered to determine what bridges would be built and maintained at the expense of the county.

With Oregon statehood in 1859, the legislature set up a system of county road districts in which supervisors and appraisers laid out county roads, assessing the cost by taxation. Dozens of requests for public roads and bridges were filed with the county courts, and their construction costs became a major portion of early county budgets. (Federal aid for Oregon's road system was scant and primarily in the form of grants for military purposes during this early period.) Many counties opted to license ferry operations rather than build bridges. With the gradual settlement of Oregon, however, bridges began to displace fords and ferries at points of heaviest traffic, but only where economic and most needed.

The first bridge in Oregon is unknown, but it may be assumed it was small, simple and undistinguished. A few log bridges were built in the Oregon Country before the migrations of the 1840s, but most were built during or after this period (Figure 9). Records indicate that a bridge was constructed on Main Street in Oregon City in the mid-1840s, followed by a second bridge to the island mills in 1847. A bridge spanning Dairy Creek in Washington County was built in 1846, and a

FIGURE 9. This timber and wrought iron queenpost truss is representative of early bridges in Oregon. Inexpensive, made of local materials, and constructed by local builders, these timber bridges were functional if not always long-lasting. (This bridge spanned Camas Creek near Ukiah, Umatilla County, and was built about 1900.)

FIGURE 10. The Booth-Kelly Railroad Bridge (also called the Hayden Bridge) across the McKenzie River at Springfield is considered the oldest surviving bridge in Oregon. The wrought iron truss structure was first erected near Corrine, Utah, in 1882 and was moved to its present location in 1900. The Whipple truss and Phoenix columns make it a unique structure.

bridge was erected across the Yamhill River at Lafayette in 1851. Subsequent bridges were built over Marys River near Corvallis in 1856, across the Tualatin River near Moore's Mill in 1859, over the Luckiamute River in 1863, and over the Long Tom in 1870. A covered timber span was built across South Mill Creek on Commercial Street in Salem in 1862, one of the first references to a covered bridge in the state.

Bridges were attempted as early as 1859 in the Rogue River section of southern Oregon. Sherar's Bridge across the Deschutes River in central Oregon was first constructed in 1858, and subsequently rebuilt several times due to periodic flooding. Timber structures were constructed across the Upper Molalla in 1861, across the Clackamas at Estacada in 1862, and across the Yamhill at Sheridan in 1865. These bridges are the earliest recorded bridges in the state.

The Railroad Era, 1870-1885

Railroad construction in the United States had a major impact on bridge building and construction. The railroads required bridges which could withstand the stress of extremely heavy loads, traveling at relatively high speeds. Certain truss forms and materials were found to perform better than others

FIGURE 11. *The first Morrison Street Bridge, a timber structure built in 1887 across the Willamette River in Portland, was replaced in 1905 by a steel bridge (shown in this 1921 photo). The swing span was open, and deck repairs were underway. The Hawthorne Bridge (1910) appears in the background. A third generation Morrison Street Bridge was completed in 1958.*

under these conditions. The railroads also required bridges which could be built cheaply and quickly. In a short period, railroad bridges proliferated and many new bridge forms were developed. Initially, many railroad bridges were timber trestles or trusses. Metal, however, was introduced in the construction of new bridges because of the structural limitations of wood and its flammability.

The 1870s saw the advent of major railroad construction in Oregon. The Oregon and California Railroad, running north and south through the state, began construction in Portland in 1868, finally reaching Ashland in 1887. The next major railroad in Oregon was the Oregon Railway and Navigation Company along the Columbia River, completed in 1883, connecting Oregon with the transcontinental system. These major lines were supplemented by short lines and interurban electric lines.

The first large bridges in Oregon were railroad spans. The Oregon and California Railroad built the first bridge across the Willamette River at Harrisburg in 1871. The structure was a covered timber Howe truss with a draw span for river navigation. The total length was 770 feet, including approaches. An equally impressive timber structure was built by the same railroad across the North Umpqua River near Roseburg in 1872. This bridge was over 1,000 feet long and consisted of five Howe trusses and trestle work. The first major steel bridge in the state was the original Steel Bridge, built across the Willamette River in Portland in 1888.

The oldest surviving bridges in Oregon are railroad structures, including the McKenzie River (Booth-Kelly or Hayden) Bridge at Springfield, built in 1882 and moved to its present location from Utah in 1900 (Figure 10). This bridge is the oldest known bridge in Oregon. Other railroad bridges of the 1880s include a span at Mill City across the North Santiam River, circa 1885, now a public footbridge; and the Armitage Railroad Bridge across the McKenzie River, north of Eugene, which serves on a bicycle route. The Armitage Bridge was built in 1887 and moved to this location in 1907. These bridges are all wrought iron truss spans.

Catalog Bridges and the Dawn of the Good Roads Movement, 1885-1900

Prior to the turn of the century, high water and flooding caused the repeated destruction of bridges and necessitated quick, inexpensive replacements. For example, catastrophic flooding occurred in 1881 and 1890, and these rampaging floods destroyed dozens of structures. Often a bridge from upstream collided with one downstream, creating a devastating, floating bridge jam. Covered bridges, because of their solid walls and buoyant timbers, were especially susceptible to washout. Fires also took their toll on wooden bridges.

FIGURE 12. Bisected by the Willamette River, Portland is a city of bridges. Ten highway bridges cross the Willamette River in Portland, seven of which are seen in this aerial view of the downtown: foreground to background, Marquam (1966), Hawthorne (1910), Morrison (1958), Burnside (1926), Steel (1912), Broadway (1913), and Fremont (1973). Not shown are St. John's (1931), Ross Island (1926) and Sellwood (1925).

FIGURE 13. *The Coos Bay Bridge at North Bend was the longest and most expensive of the five Oregon Coast bridges completed in 1936. (This view is during construction in 1935.) The Coos Bay Bridge was dedicated posthumously in 1947 to Conde B. McCullough, Oregon's early outstanding state bridge engineer.*

Destruction of an unusually large number of timber bridges in the flood of 1890 hastened a trend toward metal bridges, a trend which was already gaining momentum. Bridge building had become more complex and passed from the domain of the old-style bridge carpenter to companies specializing in light iron or steel prefabricated bridges. While the wooden bridges required frequent maintenance and were expensive to rebuild, iron and steel bridges required little maintenance and were not as apt to wash away. Sharp-talking bridge salesmen used all of their tricks to convince county courts that wooden bridges were dangerous, and therefore must be replaced. Wholesale replacement of wooden bridges followed, until the World War I period.

In Oregon, the late years of the 19th century were the heyday of the bridge companies and their catalog prefabricated bridges. The number of bridges built and the number of bridge companies mushroomed. Some companies were nationally known, although most were local or regional (Appendix B).

The late 1880s and early 1890s saw the beginning of the nationwide Good Roads movement. Initiated by bicyclists demanding smooth surfaced roads, the movement was soon joined by farmers needing better market roads and the federal government interested in good roads for rural mail deliveries. The Office of Road Inquiry was created in the United States Department of Agriculture in 1893 to investigate, educate, and distribute information on road building. (In 1916 this agency became the United States Bureau of Public Roads, the antecedent of the current Federal Highway Administration.)

Considerable interest in good roads also arose in Oregon during this period. An early hint of this concern was revealed in a Good Roads convention held in Portland in December 1896. One of the speakers defined a good road as one that was equally good in rainy weather as in fair weather, one well-drained and not in need of repair. He went on to state that no such road existed in Oregon. Subsequent county petitions to the state and federal governments indicated that good roads and bridges had a high priority in Oregon.

The 1885-1900 period saw the initial construction of large highway bridges in Oregon. After sixteen years of agitation for a highway bridge across the Willamette River in Portland, the first Morrison Street Bridge was built in 1887 (Figure 11). The Madison Street Bridge (1891) and the original Burnside Bridge (1894) were constructed a few years later. These bridges were either timber or iron trusses with swing spans to accommodate river traffic. A suspension bridge constructed at Oregon City in 1888 joined the Willamette River bridges, one of the few large suspension bridges ever built in Oregon (Figure 5).

Steel truss bridges were built across the Willamette River at Albany in 1887 and Salem in 1890. (This was Salem's second Willamette River crossing. The first was a timber bridge completed in 1886 and washed away in the big flood of 1890.) On the Oregon Coast, the first bridge over the North Fork of the Coquille River at Burton's Prairie was built in 1881. The first steel truss across the Rogue River was also built during this period at Grants Pass.

Arrival of the Motor Age and the Creation of the State Highway Department, 1900-1920

The rapid acceptance of the automobile drastically changed transportation patterns in Oregon and intensified the need for better roads and bridges. One of the results of this change was the creation of the State Highway Commission and the State Highway Department in 1913.

The State Highway Department significantly changed bridge building in the state and made rapid strides in bridge construction. By statute the new department set up a system whereby the counties could obtain bridge design services from the state at no cost. In addition, a manual on bridge specifications and designs was published in 1916, which helped to ensure a uniform standard of bridge design and contributed to high quality construction.

The United States Congress passed the Federal Aid Act in 1916, which provided funds on a matching basis to the states for road and bridge construction. To match the federal funds, Oregon adopted a gasoline tax in 1919 as a source of income for road purposes. Oregon's lead in the use of gas tax revenue for highway purposes was followed promptly by other states, and within a few years became the main source of highway revenue for all states throughout the nation.

In addition to a surge of construction, the 1900-1920 period witnessed some significant events in the bridge history of Oregon. Several major new bridges supplemented or replaced the existing Willamette River bridges in Portland, including the Hawthorne (1910), Steel (1912), and Broadway (1913) bridges (Figure 12). The Van Buren Street Bridge (1913) was built in Corvallis, and the Center Street Bridge (1918) was constructed in Salem. The largest highway bridge built during this period was the Interstate Bridge (1917) which spanned the Columbia River to connect Portland with Vancouver, Washington. (The first bridge to span the Columbia River between Oregon and Washington was the Spokane, Portland, and Seattle railroad bridge, a swing span completed in 1908.) Most of these highway bridges were steel trusses with some type of moveable span.

Several outstanding reinforced concrete structures were built during this period on the scenic Columbia River Highway in the Columbia Gorge and are among the earliest concrete bridges built by the State Highway Department. These bridges were innovative, frequently of the arch form, and beautifully sited in the Columbia Gorge environment. The bridges are now included in the Columbia River Highway Historic District, listed on the National Register of Historic Places in 1983.

Although timber truss covered bridges had been built in Oregon since the 1850s, the high point of covered bridge construction occurred at the beginning of this century, between 1905 and 1925. Oregon had a maximum of about 450 covered bridges during this period. Today, there are about 50 covered bridges remaining in the state. (A thematic nomination of Oregon's covered bridges was approved for the National Register in 1979. The bridges included in the Oregon Covered Bridges nomination, as well as in the Columbia River Highway Historic District, are presented in the historic highway bridges section of this document.)

"Pulling Oregon Out of the Mud" Era, 1920-1940"

The booming economy of the 1920s and the availability of federal-aid matching funds led to a major expansion in transportation facilities in Oregon. Higher speed cars and the increased truck traffic also necessitated changes in road and bridge design. Several major highway routes in Oregon were completed during this period. The Pacific Highway, stretching north and south from Washington to California, was dedicated in 1923. The Columbia River Highway across Oregon's northern border was paved from the coast to The Dalles in 1922. The Oregon Coast Highway, begun in 1914, was essentially finished in 1936, with the completion of five major bridges.

The design and construction of bridges in the state during this period was dominated by state bridge engineer Conde B. McCullough. McCullough's distinguished career with the State Highway Department spanned over 25 years, and he was responsible for hundreds of bridges in Oregon. He left a legacy of fine structures, and his arch bridges, in particular, have achieved the most acclaim and recognition, primarily because of their beauty. (Appendix B contains biographical information on McCullough and other significant bridge designers in Oregon.) McCullough's skill as an engineer paralleled his concern for aesthetics, and he was consistently at the forefront of new developments in bridge design. In 1931, he introduced the reinforced concrete tied arch to the United States with the construction of the Wilson River Bridge in Tillamook County. He was also the first to use the Freyssinet method of arch precompression with the Rogue River Bridge (1931) at Gold Beach.

In spite of the economic effects of the Depression, Oregon built some of its most magnificent bridges during this period with the assistance of federal PWA funding. The era of McCullough's bridge engineering culminated in the completion in 1936 of five major bridges, all crossing rivers or estuaries on the Pacific Coast and all designed to replace ferry service on the Oregon Coast Highway. The bridges—Coos Bay Bridge (North Bend), Umpqua River Bridge (Reedsport), Siuslaw River (Florence), Alsea Bay (Waldport), and Yaquina Bay (Newport)—were built at a total cost of $5.4 million and were financed in part by the Works Progress Administration. These concrete and steel bridges are impressive examples of the structures built during the Great Depression. McCullough was honored posthumously in 1947 when the Coos Bay Bridge, the largest of the five bridges, was renamed and dedicated as the McCullough Memorial Bridge (Figure 13).

Modern Era, 1940-Present

World War II limited both the funding and materials available for highway and bridge construction in Oregon. Steel was in short supply and was rarely available for bridge building. Federal and local funds were also restricted to essential projects. Oregon relied on its available resources and constructed more than half its bridges during these years from timber. (A notable exception to the timber spans is the steel, tied-arch, triple span structure across the Santiam River between Marion and Linn counties on the Pacific Highway, built in 1946.)

Highway and bridge construction accelerated after the war, but construction monies were scarce. Many of the aesthetic considerations common to earlier bridges were foregone in favor of utility and cost-effectiveness. Steel was once again available to bridge designers, and deck girder and deck truss structures were particularly common in this period.

When the Interstate Highway System was started in the late 1950s, the federal government pumped large amounts of money into road and bridge construction programs. Oregon took this opportunity to revitalize its major highway routes. Reinforced concrete was the primary bridge-building material. Steel was reserved for major structures, such as a second triple arch span over the Santiam River on Interstate 5 (1958) adjacent to the 1946 structure, the Thomas Creek Bridge on the Oregon Coast Highway (1961), and the cantilever truss across the mouth of the Columbia River at Astoria (1966). Steel remained a viable, though expensive, alternative into the 1970s, as evidenced by the Fremont Bridge (1973) across the Willamette River in Portland, a 1,255-foot arch span. (Views of post-1940 construction bridges are in Appendix F.)

The advent of prestressed and post-tensioned reinforced concrete structures, in conjunction with more stable economic conditions, brought about a revival of aesthetic interest in structures. Award-winning structures, such as the Chetco River Bridge (1972) on the Oregon Coast Highway in Curry County, and the I-205 crossing of the Columbia River (Figure 14), completed in 1982, demonstrate this new bridge aesthetic and are the latest spectacular additions to Oregon's rich bridge-building tradition.

FIGURE 14. Oregon's latest spectacular addition to its bridge-building tradition is the Interstate 205 crossing of the Columbia River, completed in 1982. Dedicated to Glenn L. Jackson, long-time member of the Oregon Transportation Commission, the twin structures are post-tensioned reinforced concrete box girders, 11,750 feet in length.

INVENTORY AND EVALUATION

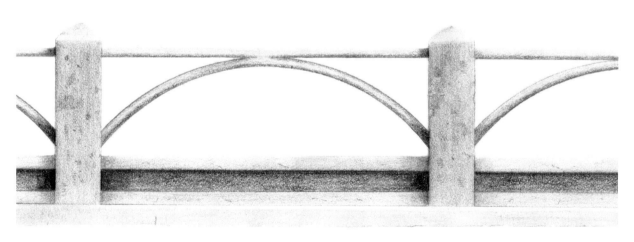

Horsetail Falls Bridge (1914), old Columbia River Highway, Oneonta vicinity, Multnomah County

INVENTORY AND EVALUATION

NEED

The primary responsibility of the Department of Transportation is to provide a safe, efficient transportation system. To finance improvements to that system, state and local funds are matched with federal funds administered by the Federal Highway Administration, United States Department of Transportation. Transportation projects using federal matching funds or requiring federal permits must be developed in accordance with the appropriate federal laws and regulations. One such body of federal law relates to historic resources and ensures that these resources are fully considered in transportation planning. The department initiated this historic highway bridges study to assist in fulfilling federal requirements on historic preservation.

The federal laws on historic preservation sometimes conflict with the department's primary responsibility. Such conflicts are not easily resolved, particularly in the case of bridge rehabilitation or replacement projects. This study, by identifying those bridges which are of historic significance, helps the department comply with the historic preservation laws. As a result, the department can better plan and schedule bridge rehabilitation and replacement projects.

In 1966, the United States Congress enacted the National Historic Preservation Act, which strengthened the federal commitment to preserving significant historic resources. While there had long been an effort to preserve historic properties of national importance, this act extended protection to historic resources with state and local significance as well.

The National Historic Preservation Act of 1966 authorized the Secretary of the Interior to develop and maintain the National Register of Historic Places. (The National Register had been created by law in 1935, but was expanded under the authority of the 1966 Act.) The National Register is the official list of districts, sites, buildings, structures, and objects significant in American history, architecture, archeology, and culture. Furthermore, the Act established the criteria for resources to be eligible for listing on the National Register (Figure 15). The National Register currently includes about 37,500 properties and districts, of which over 500 are in Oregon.

The National Historic Preservation Act of 1966 also created the Advisory Council on Historic Preservation and authorized the Council to assist in protecting National Register properties from unnecessary adverse effects due to federally funded or licensed projects (e.g., bridge replacements). Section 106 of the Act requires that projects affecting historic resources be coordinated with the Advisory Council and that the Council be given a reasonable opportunity to comment. This Act also requires that the State Historic Preservation Office be consulted and invited to comment on federally-assisted projects.

An important supplement to the 1966 Act was Executive Order 11593, issued in 1971. Previously, the law provided protection only to historic properties listed on the National Register. The Executive Order extended that same protection to all historic properties considered eligible for the National Register, regardless of whether they were listed or under consideration for listing. This order required the state to evaluate all buildings, structures, or sites potentially affected by federal projects and determine whether they are eligible for inclusion on the National Register according to the Act's criteria. If an eligible property is affected by a project, the final judgment on its eligibility is made by the Keeper of the National Register, National Park Service, United States Department of the Interior, Washington, D.C.

National Register-eligible historic properties are also afforded protection from federal projects by Section 4(f) of the Department of Transportation Act of 1966. This law states, in part, that no project can be approved for federal transportation funding that requires the use of a significant historic resource, unless there is no feasible and prudent alternative and all mitigation options have been incorporated.

Other federal or state regulations and rules to be considered include the National Environmental Protection Act (1969), the Council on Environmental Quality guidelines, the ODOT Action Plan, and Oregon's Statewide Planning Goals.

As a result of this body of legislation, historic preservation has assumed importance in highway planning. In recent years, the department has planned for the replacement or rehabilitation of a

> # NATIONAL REGISTER CRITERIA
>
> *The quality of significance in American history, architecture, archeology, and culture is present in districts, sites, buildings, structures, and objects of state and local importance that possess integrity of location, design, setting, materials, workmanship, feeling, and association and;*
>
> *(a) That are associated with events that have made a significant contribution to the broad patterns of our history; or*
> *(b) That are associated with the lives of persons significant in our past; or*
> *(c) That embody the distinctive characteristics of a type, period*
>   *or method of construction, or*
>   *Represent the works of a master, or*
>   *Possess high artistic values, or*
>   *Represent a significant and distinguishable entity whose components may lack individual distinction; or*
> *(d) That have yielded, or may likely yield, information important to prehistory or history.*
>
> *(Generally the properties must be at least 50 years old to warrant eligibility to the National Register. Properties less than 50 years can be listed, but must be of exceptional value and significance.)*
>
> FIGURE 15. *Evaluation criteria for inclusion in the National Register of Historic Places [Source: Section 101(a)(1)(A), National Historic Preservation Act of 1966, as amended, and 36 CFR Part 60.6]*

large number of bridges and has had to evaluate the historic significance of the bridges on a case-by-case basis. Comprehensive historic information about Oregon's bridge population was not available for comparative purposes. Historic judgments about bridges were difficult to make at best without knowing the characteristics of the total population of Oregon's older bridges.

PURPOSE

The purpose of the historic bridges study was to inventory, evaluate, and assess the state, county, and local highway bridges in Oregon to determine which structures were National Register eligible.

Specifically, the study serves to make bridge project development more efficient and fulfill federal laws and regulations, as follows:

1. Identifies those highway bridges which are historically significant and meet the National Register eligibility criteria.

2. Conversely, identifies bridges that are not eligible for the National Register and facilitates the replacement planning process.

3. Allows the identification of bridge replacement projects involving National Register-eligible structures early in project development, thus providing time to complete the necessary actions mandated by law.

4. Eliminates the need to conduct historic research on each bridge on a case-by-case basis.

5. Provides the comparative data to develop appropriate mitigation for historic bridges proposed for replacement.

6. Provides information for the development of a preservation priority plan.

METHODOLOGY

Basic Parameters

The Department of Transportation maintains a computerized listing of most highway bridges in state, county, and local ownerships in Oregon. This file includes about 7,000 structures and contains basic information invaluable for this study--bridge names, locations, types, dimensions, ownerships, and construction dates. This list imposed some basic parameters on the historic bridges study. A bridge was defined, thus:

> A structure including supports erected over a depression or an obstruction, as water, highway, or railway, and having a track or passageway for carrying traffic or other moving loads, and having an opening measured along the center of the roadway of more than 20 feet.

In accordance with the definition of a bridge, structures less than 20 feet were excluded from the study. Only structures of sufficient size to have involved substantial engineering were examined.

The study population consisted of highway bridges regularly inspected and potentially eligible for federal rehabilitation or replacement funding. The study did not normally include privately-owned bridges, such as railroad bridges, nor publicly-owned bridges under the jurisdictions of federal and state agencies other than the Department of Transportation. Most all local government-owned highway bridges were included. Bridges built exclusively for pedestrian use or other non-highway uses are generally excluded, except for some railroad crossings. In the course of field inspection, many of these excluded bridges were incidentally photographed and inspected, but only for comparative purposes. Other highway structures, such as tunnels, culverts, and retaining walls, were not inventoried in the study, but may be examined in future studies.

Historic resources must be at least fifty years old, as a general rule, or be of exceptional significance or value to be eligible for the National Register. The cutoff date for the construction of the bridges in this study was set at 1941, which included some bridges a few years younger than 50 years. The cutoff date was set to include Depression-era structures approaching 50 years of age. Additionally, World War II slowed domestic bridge construction, and the post-war period saw major changes in bridge building. This cutoff date enabled the study to be current for several more years and to facilitate long-range planning decisions. Application of the pre-1941 age cutoff resulted in a study population of about 1,200 highway bridges to be screened and evaluated. Several notable post-1940 bridges were also examined, but most bridges of this age group were not inventoried.

Seventy-seven highway bridges had already been determined eligible for or listed on the National Register, mostly covered bridges and structures on the old Columbia River Highway. Since their historic status was already established, these bridges did not require a full examination but were incorporated into the study.

Field Inspection and Historic Research

Field inspection and historic research were integral to the study. Field inspection verified known information and generated new data. The study team photographed each bridge and gathered important information from the nameplates, architectural and engineering details, truss configurations (where appropriate), and the setting. Bridge logs, plans, drawings, bridge maintenance records, general project files, the Oregon State Highway Commission's biennial reports, books on bridge building, professional magazines, newspapers, and historical journals provided historic information, which was supplemented with interviews with county and city public works officials.

About 500 bridges, including all of the pre-1941 truss, arch, suspension, and moveable bridges, were field inspected. Not all of the slab, beam, and girder bridges, which account for about 80 percent of the pre-1941 constructed bridges, were field checked because they were seldom historically significant. The bridges of this type were examined by random and representative sampling. A random sample of 20 percent of the slab, beam, and girder bridges was drawn, as well as these representative groups: all of the pre-1921 bridges and all the bridges in Salem, Portland, and on the Oregon Coast Highway. Instead of field inspections, some slab, beam, and girder bridges were inventoried only by an examination of the drawings and plans. These samples of the slab, beam, and girder bridges were considered adequate to isolate the historically significant bridges and to test for potential statistical occurrence in this bridge group. Including about 150 plans inventories on the slab, beam, and girder bridges, a total of about 650 bridges, or one-half of the pre-1941 bridges, were inventoried for the study. A complete list of the bridges inventoried for the study is in Appendix G.

EVALUATION AND IDENTIFICATION

Preliminary Historic Evaluation

After the field inspection and historic research stage of the study, the bridges were preliminarily evaluated for historic significance and placed in one of three categories:

Category I—National Register eligible. The inspection and research indicate the bridge is historically significant.

Category II—Possibly National Register eligible. Although of some historic importance, its suitability for the National Register awaits further information or additional age.

Category III—National Register ineligible. The bridge lacks sufficient historic significance to be considered eligible.

This preliminary evaluation resulted in 141 bridges being placed in categories I and II. The remainder of the bridges were Category III. Category I and II bridges were subjected to additional research and final evaluation.

Evaluation Rating System

The study team developed an evaluation system in consultation with the State Historic Preservation Office to identify the Oregon highway bridges considered National Register eligible. Various bridge evaluation systems developed by other states were reviewed, as well as systems used for other historic resources.

The evaluation rating system for Oregon's historic highway bridges is based on the National Register criteria with some minor modifications. To address a broad diversity of cultural resources, the National Register criteria, by necessity, are general in nature. Consequently, the criteria were tailored for specific application to bridges as a particular class of historic resources.

The evaluation rating system consisted principally of three measures of resource integrity and ten measures of historic significance, modeled after National Register criteria. (See Figure 15 for the National Register criteria.) Each measure contained a six-increment scale: unknown, none, low, moderate, high, and exceptional.

Measures of Integrity

To be eligible for the National Register, historic properties must possess integrity of location, design, setting, materials, workmanship, feeling and association. Three measures of integrity were developed to correspond to the National Register criteria on integrity. (The present structural adequacy or capacity of the bridge was not considered an integrity measure.)

Measure 1. **Location and Setting.** Location integrity is whether the property is at its original site or has been moved. Metal truss bridges, in particular, were frequently moved, as early truss fabrication and erection techniques enabled them to be conveniently relocated. Full integrity was awarded to relocated truss bridges if they had been moved to the present location over fifty years ago. Integrity was lower if the relocation occurred less 50 years ago.

Integrity of setting addresses changes to the immediate surroundings and how these changes (buildings, land use, foliage, topography, others) have affected the relationship of the property to the setting. Setting integrity was marked high if the present setting appeared to be similar to the original.

Measure 2. **Design, Materials, and Workmanship.** Integrity of design relates to whether the property retains the features of its class; that is, the essential elements of what it is intended to represent. Materials integrity considers whether original materials of historic importance have been substantially altered by deterioration or replacement and, if replaced, whether the new materials are equivalent or compatible with the original. Workmanship relates to the specific form of different materials and the way they are combined. The high rating for this measure of integrity was reserved for bridges essentially intact as constructed (except for routine maintenance). Several of Oregon's early bridges have lost some integrity, principally by the addition of new railings not compatible with the original design. Several bridges were also widened in a manner which detracted from the original integrity.

Measure 3. **Feeling and Association.** These two attributes are interpretive. Integrity of feeling and association is present if the property communicates a sense of what it was like in its particular time period. This generally occurs when the other measures of integrity are already present.

Measures of Historic Significance

The historic significance rating system included ten measures. Measures 1, 6, 7, 9, and 10 were specific elements of the National Register criteria and reflected that wording and meaning. Measures 2 through 5 were modifications of the National Register criterion "...embody the distinctive characteristics of a type, period, or method of construction." Measure 8 related to the artistic values of National Register criteria and was an attempt to measure the relationship of the bridge to the overall aesthetics of the environment.

Measure 1. **Historic Events and Persons.** This measure evaluates the bridge's association with broad patterns in Oregon's history and with famous people, exclusive of the bridge designer and names of a commemorative nature. These ratings tended to be in the none to moderate range, and when applied, were usually limited to bridges at significant crossings (Willamette River in Portland), part of a major new highway development (Oregon Coast Highway) or the Depression-era Oregon Coast Bridges project.

Measure 2. **Construction Date.** This measure rates how early the bridge was constructed in relation to Oregon's population of remaining highway bridges. Relocated truss bridges were rated on their original date of construction, not when moved to their present location. The scoring for the measure was as follows:

Built Prior to 1910 -- Exceptional
1910 - 1919 -- High
1920 - 1929 -- Moderate
1930 - 1940 -- Low
Post-1940 -- None

Measure 3. **Distinctive and Important Type.** This measure rates the singularity and distinctive quality of the bridge. A judgment on the importance of the general and specific type of bridge was included in this measure. This

43

Timber truss joint on a covered bridge (detail).

Pin connection on a steel truss (detail).

FIGURE 16. *One of the historic evaluation criteria was rarity and uniqueness. The use of obsolete and rare technologies contributes to the historic significance of bridges.*

measure rated the bridges as being representative of a category.

Measure 4. **Rarity and Uniqueness of Type.** This measure is similar to Measure 3, but basically indicates the number of bridges of each type. It also considers features and technologies, and whether they are obsolete (Figure 16). The rating for this measure included checking statistics on the bridges in the study to determine numbers represented in each type. (See Appendix D.)

Measure 5. **Engineering Innovation and Site Challenge.** This measure rates two interdependent bridge features: engineering innovations and whether the bridge was a prototype for later bridges. Certain crossings, because of width, configuration of the stream or banks, navigational needs, or subsurface conditions, posed difficult challenges for bridge building. This measure rates such difficulties and the quality of the engineering solution (Figure 17).

Measure 6. **Designer and Builder.** This measure ranks the importance of the bridge designer and builder and the importance of the bridge as it relates to the other works of that designer and builder. When the designer of the bridge was unknown and the bridge was credited to a bridge-building company, the rating was moderate or low based on the amount of information available on the particular company. (See Appendix B for significant bridge designers and bridge contractors in Oregon.)

Measure 7. **Artistic Values.** This rating measure is a judgment of the appearance

FIGURE 17. Site challenge is one of the criteria considered in the selection of the historic bridges. The Chasm Bridge (1937) is located on the near-vertical face of Neahkahnie Mountain in Tillamook County, high above the ocean on the Oregon Coast Highway.

of the bridge. This and the following measure tend toward subjective judgments. Two artistic points-of-view were considered: was there a conscious attempt to beautify the design with the addition of architectural embellishments (Figure 18), and; although unadorned, is the bridge's appearance handsome in its arrangement of functioning parts?

Measure 8. Environmental Aesthetics. This measure examines how the bridge relates to its environment and the overall aesthetics of its location. It addresses questions of scale, definition, proportion, and how well the structure complements the nearby natural and cultural landscape (Figure 19).

Measure 9. Related Historic Properties and Groupings. This measure is the National Register criterion that a property "...may be part of a group of properties that represent a significant and distinguishable entity whose components may lack individual distinction." Bridges in proximity to and historically related to other bridges or those bridges clearly part of a distinguishable and important thematic group were rated higher. Important thematic groups identified included the Willamette River bridges of Portland, Oregon's covered bridges, the reinforced concrete arch bridges of Conde B. McCullough, and the Oregon Coast Bridges project.

Measure 10. Potential for Yielding Prehistoric or Historic Information. This measure, one of the National Register criteria, usually relates to archeological properties. Its application to Oregon's bridges was rare.

45

Final Evaluation

The final identification of the National Register-eligible highway bridges was based on the evaluations of a professional review team. This team consisted of volunteer bridge experts from private companies and personnel from the State Historic Preservation Office and Oregon Department of Transportation. The team members were the following:

Lewis L. McArthur, Vice President, Ray Becker-Company, Portland

Louis F. Pierce, President and Chief Engineer, OBEC Consulting Engineers, Eugene

Thomas J. McClellan, Professor Emeritus, Oregon State University Engineering Deppartment, now a consultant with CH2M-Hill, Corvallis

David W. Powers, III, Deputy State Historic Preservation Officer, State Historic Preservation Office, Salem

Walter J. Hart, State Bridge Engineer, ODOT, Salem

Jack L. Davis, Manager, Special Studies, Bridge Section, ODOT, Salem

Pieter T. Dykman, Research Coordinator, Environmental Section, ODOT, Salem

James B. Norman, Historic Bridges Study Research Associate, Environmental Section, ODOT, Salem

Dwight A. Smith, Cultural Resources Specialist and Historic Bridges Study Manager, Environmental Section, ODOT, Salem

Formal evaluation by the review team identified 68 bridges as historically significant and National Register eligible. Fifty-three other bridges were

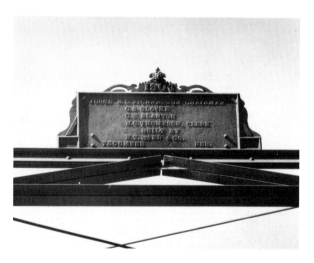

FIGURE 18. These examples illustrate the artistic embellishments on early highway bridges. The deliberate attempts to beautify bridges add interest to the structures.

identified as having some historic interest, though ineligible for the National Register. These ineligible bridges were delegated to a reserve category (Appendix E). Several notable bridges constructed after 1940 were also reviewed by the team, but were found to be ineligible for the National Register (Appendix F). The reserve and notable post-1940 bridges will be reevaluated as the study is periodically updated. Some of the bridges may become eligible in future years as National Register bridges are lost or modified and as the bridges age.

COORDINATION AND DETERMINATION OF ELIGIBILITY

The historic bridges study was developed in cooperation and consultation with the State Historic Preservation Office. In addition to participating in the design of the study, the SHPO reviewed the evaluation rating system, served on the professional review team, and reviewed draft versions of the study report. By letter on January 10, 1985, the SHPO concurred with the final list of bridges determined National Register eligible.

The owners of the eligible bridges identified in the study, including appropriate city and county public works departments, were notified of the findings of the study and the effects on federally-funded bridge replacement projects. In January 1985, the Oregon Transportation Commission approved the study report, and the study findings were released to the public.

A thematic group request for a determination of eligibility on the 68 study-identified bridges was prepared and forwarded to the Keeper of the National Register in April 1985. In May 1985, the Keeper of the National Register determined that 57 of the study-identified bridges were eligible for the

FIGURE 18 continued

FIGURE 19. Bridges relate aesthetically to the landscape. The pastoral setting of the Rhea Creek Bridge (1909) near Ruggs in Morrow County increases its appeal and significance.

National Register. (Two of the bridges in the thematic group had previously been determined eligible by the Keeper's Office--Pass Creek Covered Bridge and the Umatilla River Bridge on S.W. 10th Street.) The Keeper took no action on the remaining nine bridges in the group, structures constructed or modified less than 50 years ago, and requested supporting information that these bridges were of exceptional significance. After receiving supplemental information on the remaining nine bridges, the Keeper of the National Register determined the remaining nine bridges National Register eligible in September 1985, completing the action on the request for a determination of eligibility.

This study report is being distributed to appropriate public agencies, historic preservation organizations, libraries, and concerned citizens for their reference and information.

The study data are valid as of July 1984. In preparation for printing, additions and modifications were made, but no exhaustive attempt was made to recheck the status of all bridges.

HISTORIC HIGHWAY BRIDGES

Rogue River (Rock Point) Bridge (1920), old Pacific Highway near Gold Hill, Jackson County

HISTORIC HIGHWAY BRIDGES

This section of the document presents the historic highway bridges in Oregon. The bridges identified in the ODOT study, as well as the bridges previously determined eligible for or listed on the National Register of Historic Places, are included. Contemporary photography and descriptive information on all 145 bridges highlight this section. The historic bridges are presented in four groups:

Study Identified Bridges
Old Columbia River Highway Bridges
Covered Bridges
Other Historic Bridges

Each of these groups of historic bridges is discussed in the introductory material for each section. (A summary list of the bridges is in Figure 20.)

The 145 bridges in this section are important examples of the bridge-building tradition in Oregon and constitute a valuable legacy to the art, engineering, and technology of bridge construction. By being National Register listed or eligible, these bridges are recognized as historically significant and valuable contributors to the cultural heritage of the state.

The historic bridges illustrated in this document are distributed throughout the state. The largest proportion is in the populated urban counties of the Willamette Valley, particularly Multnomah County (31 bridges) and Lane County (22 bridges). Ten counties, many of which are in eastern Oregon, presently have no National Register bridges. (Refer to Appendix C for information on the location of the historic bridges, as well as statistics on ownership and age.)

FIGURE 20. Historic Highway Bridges of Oregon.
The historic highway bridges in Oregon determined eligible for or listed on the National Register of Historic Places are listed below in the order of presentation in this section of the study report.

STUDY IDENTIFIED BRIDGES

Truss:

Bull Run River, Clackamas County
Sandy River (Lusted Road), Clackamas County
Balch Gulch, Multnomah County
Owyhee River, Malheur County
North Fessenden Street, Multnomah County
North Willamette Boulevard, Multnomah County
Rhea Creek (Spring Hollow), Morrow County
Umatilla River (S.E. 8th Street), Umatilla County
Willow Creek (Cecil), Morrow County
Cow Creek, Malheur County
Grande Ronde River (Troy), Wallowa County
Pine Creek, Umatilla County
Siuslaw River (Richardson), Lane County
Umatilla River (S.W. 10th Street), Umatilla County
Lower Milton Creek (McDonald), Columbia County
Grande Ronde River (Old Rhinehart), Union County
Pass Creek, Douglas County
Willamette River (Albany), Linn-Benton counties
Columbia River (Bridge of the Gods), Hood River (Oregon)-Skamania (Washington) counties
Willamette River (Ross Island), Multnomah County
Santiam River (Cascadia Park), Linn County
Willamette River (Springfield), Lane County
Calapooya Creek (Rochester), Douglas County
Coos Bay (McCullough Memorial), Coos County
Crabtree Creek (Hoffman), Linn County
South Myrtle Creek (Neal Lane), Douglas County
Thomas Creek (Gilkey), Linn County

Arch:

Mill Creek (Front Street, N.E.), Marion County
Killam Creek, Tillamook County
Tumalo Irrigation Ditch, Deschutes County
Stark Street, Multnomah County
Williams Creek, Josephine County
Rogue River (Rock Point), Jackson County
N.W. Alexandra Avenue, Multnomah County

South Umpqua River (Myrtle Creek), Douglas County
Willamette River (Oregon City), Clackamas County
Necanicum River (Seaside), Clatsop County
North Umpqua River (Winchester), Douglas County
Fifteenmile Creek (Adkisson), Wasco County
Crooked River (High), Jefferson County
Depoe Bay, Lincoln County
Rocky Creek (Ben Jones), Lincoln County
Rogue River (Gold Hill), Jackson County
South Fork Hood River (Sahalie Falls), Hood River County
Cape Creek, Lane County
Rogue River (Caveman), Josephine County
Rogue River (Gold Beach), Curry County
Wilson River, Tillamook County
Clackamas River (McLoughlin), Clackamas County
Santiam River (Jefferson), Marion-Linn counties
Yaquina Bay (Newport), Lincoln County

Suspension:

Willamette River (St. John's), Multnomah County

Moveable:

Willamette River (Hawthorne), Multnomah County
Willamette River (Broadway), Multnomah County
Coquille River, Coos County
Willamette River (Burnside), Multnomah County
Siuslaw River (Florence), Lane County
Umpqua River (Reedsport), Douglas County

Slab, Beam, and Girder:

Beaver Creek (Sandy River Overflow), Multnomah County
Dollarhide, Jackson County
Old Mill Race, Umatilla County
Steinman, Jackson County
Fifteenmile Creek (Seufert), Wasco County
Mill Creek (West Sixth Street), Wasco County
Pringle Creek (Liberty Street, S.E.), Marion County
Pringle Creek/Shelton Creek (Church Street, S.E.), Marion County
Chasm (Neahkahnie Mountain), Tillamook County
Necarney Creek, Tillamook County

OLD COLUMBIA RIVER HIGHWAY BRIDGES

Sandy River (Troutdale), Multnomah County
Sandy River (Stark Street), Multnomah County
Crown Point, Multnomah County
Latourell Creek, Multnomah County
Young Creek (Shepperd's Dell), Multnomah County
Bridal Veil Falls, Multnomah County
Wahkeena Falls, Multnomah County
West Multnomah Falls, Multnomah County
Multnomah Creek, Multnomah County
East Multnomah Falls, Multnomah County
Oneonta Gorge Creek (Old), Multnomah County
Horsetail Falls, Multnomah County
Moffett Creek Bridge, Multnomah County
Tanner Creek, Multnomah County
Toothrock and Eagle Creek, Multnomah County
Eagle Creek, Multnomah County
Ruckel Creek, Hood River County
Gorton Creek, Hood River County
Ruthton Point, Hood River County
Rock Creek, Wasco County
Rock Slide Viaduct, Hood River County
Mosier Creek, Wasco County
Hog Creek Canyon (Rowena Dell), Wasco County
Chenoweth Creek, Wasco County
Dry Canyon, Wasco County
Oneonta Gorge Creek (New), Multnomah County

COVERED BRIDGES

Drift Creek, Lincoln County
Rickreall Creek (Pumping Station), Polk County
Abiqua Creek (Gallon House), Marion County
Applegate River (McKee), Jackson County
Alsea River (Hayden), Benton County
Five Rivers (Fisher School), Lincoln County
Lost Creek, Jackson County
Grave Creek, Josephine County
Mosby Creek, Lane County
Lost Creek (Parvin), Lane County
Sandy Creek (Remote), Coos County
Antelope Creek, Jackson County
Coyote Creek (Battle Creek), Lane County
Row River (Currin), Lane County
Wildcat Creek, Lane County
Yaquina River (Chitwood), Lincoln County
Evans Creek (Wimer), Jackson County

Ritner Creek, Polk County
Lake Creek (Nelson Mountain), Lane County
Elk Creek (Roaring Camp), Douglas County
Horse Creek, Lane County
Mosby Creek (Stewart), Lane County
Calapooia River (Crawfordsville), Linn County
Deadwood Creek, Lane County
Fall Creek (Unity), Lane County
Marys River (Harris), Benton County
Thomas Creek (Hannah), Linn County
Thomas Creek (Weddle), Linn County
Fall Creek (Pengra), Lane County
McKenzie River (Goodpasture), Lane County
Mill Creek (Wendling), Lane County
Mohawk River (Ernest), Lane County
North Fork Yachats River, Lincoln County
Crabtree Creek (Larwood), Linn County
North Fork of the Middle Fork Willamette River (Office), Lane County
Middle Fork Willamette River (Lowell), Lane County
South Fork Santiam River (Short), Linn County
Row River (Dorena), Lane County
Willamette Slough (Irish Bend), Benton County
South Umpqua River (Milo Academy), Douglas County
Swalley Canal (Rock O' the Range), Deschutes County
McKenzie River (Belknap), Lane County

OTHER HISTORIC BRIDGES

Rock Creek (Olex), Gilliam County
Willamette River (Steel), Multnomah County
Columbia River (Interstate), Multnomah (Oregon)-Clark (Washington) counties
Oswego Creek, Clackamas County
S.W. Vista Avenue, Multnomah County
Columbia River (Lewis and Clark), Columbia (Oregon)-Cowlitz (Washington) counties
Alsea Bay (Waldport), Lincoln County
North Umpqua River (Mott), Douglas County
Thomas Creek (Jordan), Linn County

As defined by the engineering principle used in the center or main span, the most numerous of the historic bridges are the timber trusses (covered bridges), reinforced concrete arches, and slab, beam, and girders. (Additional information on the specific structural types and examples in Oregon is found in Appendix D.)

The historic highway bridges were constructed from 1894 to 1966. Oregon has only ten historic highway bridges built before 1910. Over half (64 percent) of the historic bridges were constructed in the 1910s and 1920s. The ownerships of the historic bridges vary. The counties own the largest number (43 percent), with the Oregon Department of Transportation second, owning 38 percent of the bridges. The cities own the third largest group, 8 percent of the historic bridges. Nine of the highway bridges, mostly covered bridges, are privately owned.

STUDY IDENTIFIED BRIDGES

The 68 bridges in this group were evaluated and identified as historically significant in the ODOT historic bridges study. The thematic group consists of public highway bridges in Oregon, constructed between 1894 and 1939. Each of these bridges has undergone thorough examination, review, and research concerning the bridge's design, designer, age, technology, attention to detail, site appropriateness, and association with historic events and persons. A professional review team, consisting of ODOT personnel, the State Historic Preservation Office staff, and public volunteer members, evaluated the inventory bridges against the criteria for eligibility to the National Register.

The State Historic Preservation Office concurred that the highway bridges were National Register eligible. The bridges were subsequently determined eligible by the Keeper of the National Register as the Historic Highway Bridges of Oregon Thematic Group.

The study bridges were evaluated by comparing and contrasting individual bridges with similar bridges of the same structural type and approximate age. Accordingly, the bridges in this section are presented by structural type and are in the following order:

Truss
Arch
Suspension
Moveable
Slab, Beam, and Girder

Within these classifications, the structures are arranged chronologically by the date construction.

FIGURE 21. This perspective drawing of the Yaquina Bay Bridge at Newport, one of the study-identified bridges, was made by F.G. Hutchinson, graphics artist with the State Highway Department, in 1936.

TRUSS BRIDGES

In its simplest form, a truss is a structural frame based on the geometric rigidity of the triangle. A truss bridge consists of a framework of members forming triangles and functions as an "open" beam or cantilever. The connecting truss members of a bridge, acting in tension, compression, or both, form a rigid structure capable of supporting not only the weight of the truss, but the applied load. Truss bridges are usually of three basic designs, as illustrated at right--deck truss, pony truss, and through truss. The nomenclature for the component members of a representative through truss is shown in the glossary.

Despite Oregon's late settlement relative to much of the rest of the country, its highway system includes an impressive collection of truss bridges, ranging from 1894 wrought iron structures to nationally-recognized cantilever bridges. About 350 truss bridges are on Oregon's highway system, 217 of which were constructed prior to 1941 (Appendix D). The truss form, once a common bridge type, is rapidly disappearing due to the superior economic advantages of reinforced concrete structures.

Several distinct configurations of trusses are represented in Oregon. Examples of rare types, such as the queenpost and the half-hip Pratt trusses, are present, as well as the more common Pratt, Parker and Warren configurations.

Rare and obsolete bridge construction techniques are evident in the 25 pin-connected highway truss bridges revealed by the study. Eleven of the 25 pin-connected trusses are National Register-eligible bridges and are shown on the following pages. (Nine pin-connected trusses are reserve bridge in Appendix E.) These early pin-connected metal trusses occasionally exhibit wrought iron

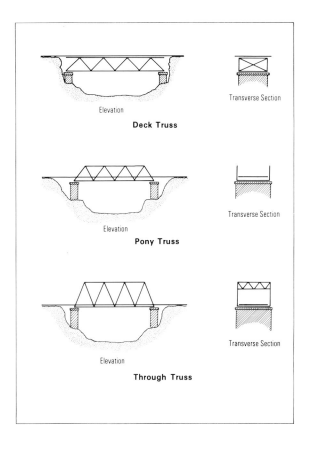

members, portal messages (unique to the "catalog" prefabricated bridges), and ornate lattice railings. The riveted truss technology displaced pin connections during the early part of the 20th century and now forms the bulk of the truss population. The riveted trusses also include some rare and interesting forms, including the double-intersection Warren.

Twenty-seven truss structures were identified as significant and are presented in the following pages.

57

THE TRUSS BRIDGES

Bridge Name	Bridge Number	County	Type	Date
Bull Run River	6571	Clackamas	Iron Through Truss* (Pennsylvania-Petit)	1894 (Site-1926)
Sandy River (Lusted Road)	6580	Clackamas	Iron Through Truss* (Pennsylvania-Petit)	1894 (Site-1926)
Balch Gulch	25B15	Multnomah	Steel Deck Truss* (Pratt)	1905
Owyhee River	45C611	Malheur	Steel Through Truss* (Parker)	1906
North Fessenden Street Overcrossing	6.7	Multnomah	Steel Deck Truss (Warren)	1909
North Willamette Boulevard Overcrossing	5.8	Multnomah	Steel Deck Truss (Warren)	1909
Rhea Creek (Spring Hollow)	49C05	Morrow	Steel Through Truss* (Pratt)	1909
Umatilla River (S.E. 8th Street)	24T01	Umatilla	Steel Through Truss* (Pratt)	1909
Willow Creek (Cecil)	10949	Morrow	Steel Pony Truss* (Half-Hip Pratt)	1909
Cow Creek	45C609	Malheur	Steel Through Truss* (Pratt)	Ca. 1910
Grande Ronde River (Troy)	32C362	Wallowa	Steel Through Truss* (Pennsylvania-Petit)	1910
Pine Creek	59C534	Umatilla	Steel Pony Truss* (Queenpost)	Ca. 1910 (Site-1935)
Siuslaw River (Richardson)	39C501	Lane	Steel Through Truss (Double-Intersection Warren)	1912 (Site-1956)
Umatilla River (S.W. 10th Street)	24T03	Umatilla	Steel Pony Truss (Warren with Polygonal Upper Chords)	1913
Lower Milton Creek (McDonald)	13757	Columbia	Steel Pony Truss (Warren)	1914
Grande Ronde River (Old Rhinehart)	799	Union	Steel Deck Truss (Warren with Verticals)	1922
Pass Creek	19B01	Douglas	Timber Through Truss (Howe) Covered Bridge	1925
Willamette River (Albany)	1025D	Linn-Benton	Steel Through Truss (Parker)	1925
Columbia River (Bridge of the Gods)	—	Hood River	Steel Through Truss (Cantilever)	1926
Willamette River (Ross Island)	5054	Multnomah	Steel Deck Truss (Cantilever)	1926
Santiam River (Cascadia Park)	1356	Linn	Timber Deck Truss (Howe)	1928
Willamette River (Springfield)	1223	Lane	Steel Through Truss (Warren with Polygonal Upper Chords)	1929
Calapooya Creek (Rochester)	1893	Douglas	Timber Through Truss (Howe) Covered Bridge	1933
Coos Bay (McCullough Memorial)	1823	Coos	Steel Through Truss (Cantilever)	1936
Crabtree Creek (Hoffman)	1724	Linn	Timber Through Truss (Howe) Covered Bridge	1936
South Myrtle Creek (Neal Lane)	10C220	Douglas	Timber Through Truss (Kingpost) Covered Bridge	1939
Thomas Creek (Gilkey)	12943	Linn	Timber Through Truss (Howe) Covered Bridge	1939

*Pin-connected bridge. The Rock Creek (Olex) Bridge in the Other Historic Bridges section is also pin connected. Additional pin-connected truss bridges are shown in Appendix E (Reserve Bridges).

Structure Number 6571
Bull Run County Road
Bull Run, Clackamas County

Constructed - 1894 (Site-1926)
Iron Through Truss (Pennsylvania-Petit)
Ownership - Clackamas County

BULL RUN RIVER BRIDGE

The Bull Run River Bridge, pin-connected and incorporating both wrought iron and steel structural members, is of major significance to Oregon's bridge heritage. The use of wrought iron predates 1900 and represents a now obsolete material. The Bull Run River Bridge is one of only two bridges of its type on Oregon's highway system. The other is the Sandy River Bridge on Lusted Road, also in Clackamas County. The Bull Run River Bridge contains a Pennsylvania-Petit truss, 240 feet in length. The span was originally part of the Burnside Bridge across the Willamette River in Portland, built in 1894 by the Bullen Bridge Company. The truss portals contain nautical design elements, appropriate to its former location. The bridge was relocated to its present site in the Bull Run Reservoir area when the current Burnside Bridge was built in 1926.

Structure Number 6580
Lusted County Road
Sandy vicinity, Clackamas County

Constructed - 1894 (Site-1926)
Iron Through Truss (Pennsylvania-Petit)
Ownership - Clackamas County

SANDY RIVER (LUSTED ROAD) BRIDGE

One of only two remaining pin-connected trusses utilizing wrought iron members on Oregon's highway system, the Sandy River Bridge is an important example of the technology of a bygone era of bridge building. The structure exhibits ornamentation virtually unknown on later truss bridges. The entrance portals of this 300-foot Pennsylvania-Petit truss have miniature ship's wheels and decorative flanges, derived from the structure's original location as part of the Burnside Bridge built in 1894 across the Willamette River in Portland. When the current Burnside Bridge was constructed in 1926, this is one of the two spans of the original bridge relocated to Clackamas County. The Sandy River Bridge is located in Dodge Park, a Clackamas County park. This bridge is located adjacent to a similar pin-connected wrought iron truss carrying water lines, built by the Bullen Bridge Company in 1893.

Structure Number 25B15
N.W. Thurman Street, MacLeay Park
Portland, Multnomah County

Constructed - 1905
Steel Deck Truss (Pratt)
Ownership - City of Portland

BALCH GULCH BRIDGE

This 160-foot Pratt truss is the oldest highway deck truss in Oregon and is one of only two remaining pin-connected deck trusses. The Balch Gulch Bridge is one of only four hanging truss designs in the state. Designed by J.B.C. Lockwood, a consulting engineer in Portland, the bridge was completed in 1905 at a cost of $35,000 and replaced a timber structure. The upgrading of N.W. Thurman Street, including the construction of this bridge, was completed in 1905 in preparation for the Lewis and Clark Exposition held nearby at the Guild's Lake area. The structure is supported by steel trestles and overlooks MacLeay Park.

Structure Number 45C611
County Road
Rome vicinity, Malheur County

Constructed - 1906
Steel Through Truss (Parker)
Ownership - Malheur County

OWYHEE RIVER BRIDGE

The Owyhee River Bridge is the oldest through truss bridge at its original location on Oregon's highway system and is the fourth oldest highway bridge in the state. Built by the H.T. Ward Company of Tecumseh, Nebraska, the structure is pin connected and has a portal message, an ornamental plaque mounted above the entrance which gives the construction date, names of local officials, and the bridge builder. The structure, a 152-foot steel through Parker truss, is supported by concrete-filled steel caissons. A 40-foot pin- connected steel pony truss of the half-hip Pratt type serves as a secondary span. (There are only three known pony trusses of this type in Oregon.)

Structure Number 6.7
North Fessenden Street
Portland, Multnomah County

Constructed - 1909
Steel Deck Truss (Warren)
Ownership - Burlington Northern Railroad

NORTH FESSENDEN STREET OVERCROSSING

This 90-foot hanging steel deck truss contains a Warren truss and is one of four highway structures built in 1909 by the Spokane, Portland, and Seattle Railroad in North Portland's Peninsula district across the Portsmouth Cut. The sidewalk railings are an early design of lattice woven steel. The riveted connections in the truss herald the dawn of this innovation, which rendered the older pin-connected technology obsolete by about 1915. The North Fessenden Overcrossing is one of the oldest extant riveted truss bridges on Oregon's highway system. This street viaduct is a part of a 4-3/4-mile long complex of railroad bridges, causeways, and cut from the north bank of the Columbia River to the south bank of the Willamette River. The railroad bridges in this complex are comprised of 21 through truss spans, three center swing drawspans and three deck plate girder spans, a total of 1,767 feet of structures, constructed between 1906 and 1908. All of the structures, including the highway bridges across the cut, were designed by internationally-known bridge engineer Ralph Modjeski.

Structure Number 5.8
North Willamette Boulevard
Portland, Multnomah County

Constructed - 1909
Steel Deck Truss (Warren)
Ownership - Burlington Northern Railroad

NORTH WILLAMETTE BOULEVARD OVERCROSSING

This highway structure has three 90-foot hanging Warren truss spans supported by steel towers. Built in 1909 by the Spokane, Portland, and Seattle Railroad, it is distinctive as one of only four hanging deck trusses in the state. Of additional interest is the vintage lattice steel railing. The North Willamette Overcrossing is also one of the earliest examples of a riveted truss in Oregon. This structure was designed by prominent bridge engineer Ralph Modjeski. (Additional information about the Spokane, Portland, and Seattle Railroad project in North Portland is in the description of the North Fessenden Street Overcrossing.)

Structure Number 49C05 *Constructed - 1909*
County Road 705 *Steel Through Truss (Pratt)*
Ruggs vicinity, Morrow County *Ownership - Morrow County*

RHEA CREEK (SPRING HOLLOW) BRIDGE

Located in a rural environment amid the rolling rangeland of eastern Oregon, the Rhea Creek Bridge is a notable example of pin-connected bridges. The portal messages, a feature which appears on only a few bridges in Oregon, are significant details. The portal message gives the name of the builder (Columbia Bridge Company of Walla Walla) and the date of construction, while a name plaque lists county officials from the period. The 114-foot steel truss is a Pratt type.

Structure Number 24T01
S.E. 8th Street
Pendleton, Umatilla County

Constructed - 1909
Steel Through Truss (Pratt)
Ownership - City of Pendleton and Umatilla County

UMATILLA RIVER (S.E. 8TH STREET) BRIDGE

The Umatilla River Bridge was constructed by the Columbia Bridge Company of Walla Walla, as indicated on the portal messages. The use of pin connections contributes to the significance of this bridge. Only twenty-five pin-connected trusses remain on Oregon's highway system. The structure consists of two 120-foot steel spans of Pratt configuration. The bridge was structurally modified in 1951 to increase its load capacity, but the rehabilitation work was integrated into the design and does not detract from its appearance or function.

Structure Number 10949
Cecil County Road 546 *
Cecil, Morrow County

Constructed - 1909
Steel Pony Truss (Half-Hip Pratt)
Ownership - Morrow County

WILLOW CREEK (CECIL) BRIDGE *

The Willow Creek Bridge is a 60-foot steel pony truss, built by the Columbia Bridge Company of Portland. The pin connections and the rarity of the truss type, a half-hip Pratt, contribute to the significance of this bridge. The original 1909 name plaque is intact, and the roadway deck is of timber construction. The setting in the hilly rangeland of eastern Oregon is pleasant, and the unpaved country road is lightly traveled, adding to the historic feel of this early truss. Cecil, a small community which had a railroad station and post office, derived its name from the Cecil family, landowners in the area. The Oregon Trail crossed Willow Creek in the vicinity of the bridge.

* This bridge was relocated to a nearby ranch in 1987.

Structure Number 45C609
County Road 793
Danner vicinity, Malheur County

Constructed - Ca. 1910
Steel Through Truss (Pratt)
Ownership - Malheur County

COW CREEK BRIDGE

This 79-foot Pratt truss is pin-connected and typical of the metal trusses at the turn of the century. The Cow Creek Bridge achieves significance not only as a representative of pin-connected technology, but also due to its association with a significant early transportation route and nearby historic structures. A 1910 photo of the nearby Inskip Station shows the bridge in place. The bridge is on the Central Oregon Military Wagon Road (established in 1867) which became part of the Idaho-Oregon-Nevada (I.O.N.) Highway around 1881.

Structure Number 32C362 Constructed - 1910
County Road Steel Through Truss (Pennsylvania-Petit)
Troy, Wallowa County Ownership - Wallowa County

GRANDE RONDE RIVER (TROY) BRIDGE

The Troy Bridge stands among Oregon's early truss bridges as one of the most attractive and well-preserved examples of pin-connected technology. The main span of the Troy Bridge is a 175-foot long, pin-connected, steel truss of Pennsylvania-Petit configuration. A 60-foot steel Warren pony truss serves as a secondary span. Name plaques and portal messages list several local officials, the construction date, and the bridge builder, the Columbia Bridge Company of Walla Walla.

Structure Number 59C534 Constructed - Ca. 1910 (Site-1935)
County Road 697 Steel Pony Truss (Queenpost)
Umapine vicinity, Umatilla County Ownership - Umatilla County

PINE CREEK BRIDGE

Though no nameplate is present on this structure and historic information is sketchy, the use of pin-connected technology establishes the construction date at about 1910. Records indicate the bridge was moved to its current location about 1935, where it serves a lightly traveled county road in the farmland of northeast Oregon. The roadway deck of the 51-foot steel pony truss is timber. The Pine Creek Bridge is important both for its pin connections and its rare truss type. This queenpost truss is one of only four in Oregon.

Structure Number 39C501
Richardson County Road 1288 *
Richardson, Lane County

Constructed - 1912 (Site-1956)
Steel Through Truss (Double-Intersection Warren)
Ownership - Lane County

SIUSLAW RIVER (RICHARDSON) BRIDGE *

Built by the Coast Bridge Company in 1912, this structure is a 124-foot, double-intersection Warren, steel through truss. The Richardson Bridge is one of only two through trusses in Oregon with this truss type and is the oldest remaining through truss connected with rivets instead of pins. The other double-intersection Warren truss is the Crooked River (Elliott Lane) Bridge, Ca. 1914, in Crook County. (See Appendix E.) Originally located across the Crooked River near Prineville, the Richardson Bridge was moved to its current site in 1956. The bridge replaced an earlier nearby timber bridge. The community of Richardson was named for the pioneer family who settled in the area in 1889. Richardson family members still reside near the bridge.

* This bridge was relocated to Eugene for pedestrian use in 1987.

Structure Number 24T03
S.W. 10th Street *
Pendleton, Umatilla County

Constructed - 1913
Steel Pony Truss (Warren with Polygonal Upper Chords)
Ownership - City of Pendleton

UMATILLA RIVER (S.W. 10TH STREET) BRIDGE *
(Also, Star Street Bridge)

This 311-foot structure consists of three steel pony trusses, Warren trusses with polygonal upper chords. It is one of Oregon's earliest highway trusses built using rivet connections and illustrates the concept of utilizing polygonal, rather than straight, top chords. This design reduces the dead load of the structure and the amount of materials needed. The bridge has a lattice steel railing. It is the longest of Oregon's pre-1941 pony trusses and the only one to utilize three pony spans in series. The S.W. 10th Street Bridge was historically known as the Star Street Bridge. The name of the street was changed in the 1940-50s period. The bridge drawings indicate it was built by the American Bridge Company of New York. This is the only known bridge in Oregon designed by that prolific company, started originally by Andrew Carnegie. This bridge was determined eligible for the National Register in April 1985.

* This bridge was removed in 1987 and placed in storage for possible reuse.

Structure Number 13757　　Constructed - 1914
Milton Way　　Steel Pony Truss (Warren)
St. Helens, Columbia County　　Ownership - City of St. Helens

LOWER MILTON CREEK (McDONALD) BRIDGE

This 85-foot, rivet-connected, steel pony truss is one of the early bridges designed and constructed by the newly formed (1913) Oregon State Highway Department. The name "Henry L. Bowlby," Oregon's first state highway engineer, and the date "1914" appear on the name plaque, as well as the bridge number "11." The bridge was designed by C.H. Purcell, the first state bridge engineer. The truss type is a Warren, a relatively common type. The roadway deck is steel grating, which replaced the original creosoted timber deck in the 1930s. The structure was built by the Ambrose-Burdsall Company and completed in December 1914. A similar truss was also built in 1914 across Upper Milton Creek. The Lower Milton Creek Bridge served originally on the Pittsburg-St. Helens Highway.

Structure Number 799
Wallowa Lake Highway 10, M.P. 16.43 (Bypassed)
Elgin vicinity, Union County

Constructed - 1922
Steel Deck Truss (Warren with Verticals)
Ownership - State of Oregon

GRANDE RONDE RIVER (OLD RHINEHART) BRIDGE

The Old Rhinehart Bridge is a 320-foot structure, with a 142-foot main span containing a steel deck truss, Warren with verticals. The structure has been bypassed and abandoned. The remote setting and abandoned condition emphasize the historic feeling of the old bridge. The Old Rhinehart Bridge was constructed under the auspices of Conde B. McCullough, Oregon's noted state bridge engineer, and is a fine example of early steel deck truss construction in Oregon. The ornamental railing, support piers, brackets, and girder members are all essentially standard designs of the period.

Structure Number 19B01 *Constructed - 1925*
First Street * *Timber Through Truss (Howe) Covered Bridge*
Drain, Douglas County *Ownership - City of Drain*

PASS CREEK BRIDGE *

The Pass Creek Bridge at Drain is a 61-foot housed Howe truss. The structure is one of the few Oregon covered bridges within city boundaries and is only two blocks from the Drain city center. The City of Drain owns the bridge. The covered bridge parallels a 1906 steel truss railroad bridge. Though the structure has been closed to traffic for many years, children still use the bridge on their way to and from a nearby school. The structure is relatively void of architectural detailing and has no openings for light except the portals. A marker indicates the original 1870s bridge at this site served as part of the overland stage route from Roseburg to Scottsburg. The bridge was determined eligible for the National Register in February 1985.

* The Pass Creek Bridge was moved upstream to a small park in 1988.

Structure Number 1025D
Albany-Corvallis Highway 31, M.P. 10.44
Albany-North Albany, Linn-Benton counties

Constructed - 1925
Steel Through Truss (Parker)
Ownership - State of Oregon

WILLAMETTE RIVER (ALBANY) BRIDGE

An ornate railing and entrance pylons adorn this 1,090-foot structure over the Willamette River at Albany. The main spans of the bridge are four 200-foot steel Parker trusses. The bridge was designed by the state bridge engineer Conde B. McCullough and was constructed by the Union Bridge Company of Portland. The Albany Bridge is one of only a few remaining multiple truss spans in Oregon. This structure replaced an 1887 bridge across the Willamette River.

Structure Unnumbered
Bridge of the Gods Road
Cascade Locks, Hood River County (Oregon)
and Stevenson vicinity, Skamania County (Washington)

Constructed - 1926
Steel Through Truss (Cantilever)
Ownership - Port of Cascade Locks

COLUMBIA RIVER (BRIDGE OF THE GODS)

The Bridge of the Gods is a 1,856-foot long structure. The 1,131-foot steel cantilever through truss has a main span of 706 feet. Built by the Wauna Toll Bridge Company of Walla Walla, the original bridge was 1,127 feet long. When the Bonneville Dam was constructed, the structure was raised and lengthened to accommodate the rising water level. The bridge is significant not only as a fine example of cantilever technology and as a major crossing of the Columbia River, but also because of its location in the Columbia River Gorge. The bridge takes its name from an Indian myth describing a large natural rock bridge over the Columbia River at this site. The Bridge of the Gods is the third oldest highway bridge spanning the Columbia River along the Oregon-Washington border.

Structure Number 5054
Mt. Hood Highway 26, M.P. 0.77
Portland, Multnomah County

Constructed - 1926
Steel Deck Truss (Cantilever)
Ownership - State of Oregon

WILLAMETTE RIVER (ROSS ISLAND) BRIDGE

The Ross Island Bridge is the only cantilever deck truss in Oregon. The structure appears to be a steel deck arch though it functions structurally as a cantilever truss. It was designed by internationally recognized consulting engineer Gustav Lindenthal. The main span was constructed by Booth and Pomeroy, while the approaches were contracted to Lindstrom and Feigenson. The 1,819-foot long cantilever truss contains a 535-foot main span. Including the twenty-nine approach spans, the length of the total structure is over 3,700 feet. In addition to its unique structural design, the spindle-type balustrade railing contributes to the appeal of this bridge. The Ross Island Bridge is one of ten highway bridges across the Willamette River in Portland and is the most southerly one in the central business district. It is one of three Willamette River bridges built by Multnomah County in the mid-1920s (Sellwood, 1925, and Burnside, 1926).

Structure Number 1356
Cascadia State Park Road
Santiam Highway 16, M.P. 41.41
Cascadia, Linn County

Constructed - 1928
Timber Deck Truss (Howe)
Ownership - State of Oregon

SANTIAM RIVER (CASCADIA PARK) BRIDGE

This structure is the only timber deck truss remaining on Oregon's highway system. The bridge is a 120-foot Howe truss span and is the only known Howe truss in Oregon that is not housed or covered. The roadway partially protects the truss from the elements. Though the steel vertical tension members are original, the timber portions of the truss have been replaced. The bridge plans are signed by Conde B. McCullough, State Bridge Engineer. The bridge was designed for Linn County by the State Highway Department, but was constructed by county forces. The bridge is located at the entrance to Cascadia State Park and is now state owned.

Structure Number 1223
McKenzie Highway 15, M.P. 1.33
Springfield, Lane County

Constructed - 1929
Steel Through Truss (Warren with Polygonal Upper Chords)
Ownership - State of Oregon

WILLAMETTE RIVER (SPRINGFIELD) BRIDGE

The Springfield Bridge is a 1,090-foot structure, with a 550-foot steel continuous through truss main span. The bridge was designed by Conde B. McCullough, evidenced by the ornate entrance pylons and decorative concrete railing. (McCullough-designed structures usually exhibit a variety of architectural treatments, a popular nationwide style in the 1920-30s.) The Springfield Bridge is one of very few truss bridges to receive such artistic treatment. In addition, it is the largest non-cantilever truss span in the state and one of only three pre-1941 continuous truss designs.

Structure Number 1893
County Road 10A
Sutherlin, Douglas County

Constructed - 1933
Timber Through Truss (Howe) Covered Bridge
Ownership - Douglas County

CALAPOOYA CREEK (ROCHESTER) BRIDGE

This 80-foot housed Howe truss span was built by veteran Oregon covered bridge builder Floyd C. Frear. The Rochester Bridge is an important representative of covered bridge design and is well suited to its rural setting. The design of this bridge is unique among Oregon's housed structures, featuring side windows having graceful curved tops. There are four openings on either side. The portals have flat arched openings. The exposed false beams at the gable ends add architectural interest as well.

Structure Number 1823
Oregon Coast Highway 9, M.P. 234.03
North Bend, Coos County

Constructed - 1936
Steel Through Truss (Cantilever)
Ownership - State of Oregon

COOS BAY (McCULLOUGH MEMORIAL) BRIDGE

Dedicated posthumously in 1947 to its designer, Conde B. McCullough, this cantilever truss exhibits an array of architectural and decorative features. In addition to its distinction as one of the most impressive of Oregon's bridges, it was the longest structure on Oregon's highway system when constructed, 5,305 feet. The 1,709 foot through truss has a main span of 793 feet and is flanked by thirteen open-spandrel, rib-type reinforced concrete deck arches. To ease design conflict between the steel truss and the arch spans, the cantilever was constructed with curved upper and lower chords. David Plowden in **Bridges: The Spans of North America**(1974) states, "Few later bridges of its type have been as outstanding." This bridge was the largest of five bridges constructed as part of the federally-assisted Oregon Coast Bridges project that replaced ferry service on major Oregon rivers and estuaries and essentially completed the Oregon Coast Highway (U.S. 101).

Structure Number 1724
Hungry Hill Drive County Road 647
Crabtree vicinity, Linn County

Constructed - 1936
Timber Through Truss (Howe) Covered Bridge
Ownership - Linn County

CRABTREE CREEK (HOFFMAN) BRIDGE

The Hoffman Bridge, a 90-foot housed Howe truss, was built by Lee Hoffman in 1936 following State Highway Department standardized plans. To build this structure, trees were cut on nearby Hungry Hill and hauled by horses to the construction site. The structure is an attractive and well-crafted example of the covered bridge tradition in Oregon. The bridge has small gothic-style windows. It is located in a rural environment near Crabtree, settled in 1845. The portal design, originally rounded when the bridge was constructed, was enlarged and squared to permit larger loads. This bridge was added to the National Register in February 1987.

Structure Number 10C220 Constructed - 1939
Neal Lane County Road 124 Timber Through Truss (Kingpost) Covered Bridge
Myrtle Creek vicinity, Douglas County Ownership - Douglas County

SOUTH MYRTLE CREEK (NEAL LANE) BRIDGE

One of the shortest covered bridges in Oregon at 42 feet, the Neal Lane Bridge is the only covered bridge with a kingpost truss in the state. The plank flooring, arched portals, narrow window openings, and rural setting add to the appeal of this bridge. The Neal Lane Bridge, built for only $1,000, was constructed by Douglas County in 1939. Floyd C. Frear was the county engineer, with Homer Gallop, the bridge foreman.

Structure Number 12943
Goar County Road 629
Scio vicinity, Linn County

Constructed - 1939
Timber Through Truss (Howe) Covered Bridge
Ownership - Linn County

THOMAS CREEK (GILKEY) BRIDGE

This distinctive representative of Oregon's covered bridge heritage takes its name from the community of Gilkey. The community, established in 1880, was a center for shipment of farm products, but has since virtually disappeared leaving only the covered bridge as a reminder of its existence. The structure is a 120-foot housed Howe truss. The large side openings are distinctive of seven similar covered bridges in Linn County. Until 1960, a covered railroad bridge stood adjacent to the Gilkey Bridge. This bridge was listed on the National Register in February 1987.

ARCH BRIDGES

An arch bridge is a bridge type in which convexly curved structural members span an opening and provide support for the roadway. Load stresses are transferred to piers or abutments through direct compression of the structural members. Arch bridges are usually of two basic forms, a deck arch and a through arch. Half-through arches, however, are also present with the roadway crossing through the middle of the arch. Deck arches are either open or filled spandrel in design, the spandrel being the area above the arch and below the roadway.

Arch bridges are particularly adapted to architectural treatment, as demonstrated superbly in Oregon's highly detailed arch structures. Unlike the other bridge types, arch structures are typically ornate. It is the arch bridges which seem to generate recognition and local pride. Where the truss can evoke nostalgia from the engineer, it is invariably the arch structures which seem able to stimulate such feelings in the remainder of the community.

Oregon currently has 64 arch bridges on its highway system. Fifty-three of these structures were built prior to 1941, with some dating back to about 1910. Many of the earliest arches have been structurally modified and no longer retain their original integrity.

The arch bridges designed by Conde B. McCullough comprise a special class of bridges in Oregon. Oregon's noted state bridge engineer created innovative and artistic structures, many of which have gained national attention. In his long career with the State Highway Department, he is credited with designing and constructing 34 major arch structures.* Only two of his arch bridges have been replaced: the Keno Bridge (1931), a three-span through reinforced concrete structure across the Klamath River in Klamath County; and the Eagle Creek Bridge, a three-span steel tied arch (1936) on the Columbia River Highway in Multnomah County. One span of the Eagle Creek

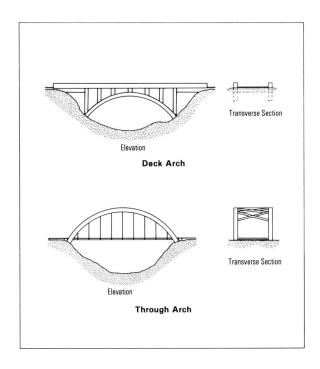

bridge was saved and is now the Clackamas River (Barton) Bridge (1970), Clackamas County.

Many of McCullough's arch bridges have been modified over the years, and several others have now reached the end of their useful service and are scheduled for replacement. This document includes views and descriptions of 32 McCullough arches, including 19 study identified arch bridges. Four other McCullough arch bridges have previously been listed on the National Register or determined eligible and are shown in the Old Columbia River Highway Bridges and Other Historic Bridges sections of this document. Nine of his arch structures are in the Reserve Bridges section (Appendix E).

Twenty-four arch bridges (including both McCullough and non-McCullough designs) were identified as historically significant structures and are presented on the following pages.

---

*Several McCullough bridges consist of non-arch mainspans, but have arched secondary spans. These bridges are included in the arch bridges count above, but are located in the document under the main span type in the Study Identified Bridges section. They include the Coos Bay (McCullough Memorial) Bridge--steel cantilever truss, the Siuslaw River (Florence) Bridge--steel bascule span, and the Umpqua River (Reedsport) Bridge--steel swing span.

THE ARCH BRIDGES

Bridge Name	Bridge Number	County	Type	Date
Mill Creek (Front Street N.E.)	13811A	Marion	Reinforced Concrete Deck Arch	1913
Killam Creek	57C29	Tillamook	Reinforced Concrete Deck Arch	1914
Tumalo Irrigation Ditch	17C02	Deschutes	Reinforced Concrete Deck Arch	1914
Stark Street	11113	Multnomah	Reinforced Concrete Deck Arch	1915
Williams Creek	2379	Josephine	Reinforced Concrete Deck Arch	1917
Rogue River (Rock Point)	332A	Jackson	Reinforced Concrete Deck Arch	1920
N.W. Alexandra Avenue	25B14	Multnomah	Reinforced Concrete Deck Arch	1922
South Umpqua River (Myrtle Creek)	490A	Douglas	Reinforced Concrete Deck Arch	1922
Willamette River (Oregon City)	357	Clackamas	Steel Half-Through Arch	1922
Necanicum River (Seaside)	7C11	Clatsop	Reinforced Concrete Deck Arch	1924
North Umpqua River (Winchester)	839	Douglas	Reinforced Concrete Deck Arch	1924
Fifteenmile Creek (Adkisson)	1095	Wasco	Reinforced Concrete Deck Arch	1925
Crooked River (High)	600	Jefferson	Steel Deck Arch	1926
Depoe Bay	2459	Lincoln	Reinforced Concrete Deck Arch	1927
Rocky Creek (Ben Jones)	1089	Lincoln	Reinforced Concrete Deck Arch	1927
Rogue River (Gold Hill)	576	Jackson	Reinforced Concrete Deck Arch	1927
South Fork Hood River (Sahalie Falls)	3545-000-1.2	Hood River	Reinforced Concrete Deck Arch	1928
Rogue River (Caveman)	1418	Josephine	Reinforced Concrete Half-Through Arch	1931
Rogue River (Gold Beach)	1172	Curry	Reinforced Concrete Deck Arch	1931
Wilson River	1499	Tillamook	Reinforced Concrete Through Tied Arch	1931
Cape Creek	1113	Lane	Reinforced Concrete Deck Arch	1932
Clackamas River (McLoughlin)	1617	Clackamas	Steel Through Tied Arch	1933
Santiam River (Jefferson)	1582	Marion/Linn	Reinforced Concrete Through Arch	1933
Yaquina Bay (Newport)	1820	Lincoln	Steel Half-Through Arch	1936

Structure Number 13811A
Front Street, N.E.
Salem, Marion County

Constructed - 1913
Reinforced Concrete Deck Arch
Ownership - City of Salem

MILL CREEK (FRONT STREET, N.E.) BRIDGE

This 40-foot filled spandrel concrete barrel arch is the oldest unaltered arch structure on Oregon's highway system. The bridge was originally a railroad bridge built for the Oregon Electric Railway by Hurley-Mason and Company and designed by A. M. Lupfer of the Spokane, Portland, and Seattle Railway. The date of construction, 1913, is shown at the top of the arch. The structure functions as a railroad bridge, but also provides automobile and truck traffic service.

Structure Number 57C29
South Prairie County Road 765
Tillamook vicinity, Tillamook County

Constructed - 1914
Reinforced Concrete Deck Arch
Ownership - Tillamook County

KILLAM CREEK BRIDGE

The Killam Creek Bridge is a 52-foot multi-radius concrete deck arch designed by U.G. Jackson, the county surveyor for Tillamook County. The structure is one of the oldest extant arch bridges in the state and is a typical example of the early applications of reinforced concrete to short-span highway bridges. The bridge has little architectural or decorative treatment. The low railings are filled concrete with some fluting, and two of the four original railing end posts are missing. The names of county officials from the period are listed on the bridge railings. The structure was built by local contractor, C. F. Feldschau, for $2,500. The Killam Creek Bridge was on the Oregon Coast Highway prior to the relocation of the highway to a new alignment in the late 1940s.

Structure Number 17C02
Seisemore County Road
Tumalo vicinity, Deschutes County

Constructed - 1914
Reinforced Concrete Deck Arch
Ownership - Deschutes County

TUMALO IRRIGATION DITCH BRIDGE
(Also, Bull Creek Dam Bridge)

The Tumalo Irrigation Ditch Bridge (historically, Bull Creek Dam Bridge) was constructed as part of the ill-fated Tumalo Irrigation Project (1902-15). The Tumalo Bridge is notable for its role in this state reclamation project and for its unique design. The bridge consists of five continuous filled-spandrel, barrel-type, concrete deck arch spans, each 25 feet long. The concrete piers are keyed into notches in the arch structure. Four panels on the spandrel walls read, "State of Oregon 1914." The bridge, dam and spillway were designed by Olaf Laurgaard, who later became city engineer for Portland. The Bull Creek Dam under the bridge and the nearby Tumalo Dam were constructed to form a water storage reservoir to increase the amount of irrigated acreage at Tumalo. The Bend-to-Sisters road was rerouted around the eastern perimeter of the reservoir, crossing both dams. (This is the present-day, unpaved Seisemore Road.) When water was turned into the basin after dam construction, it failed to hold the storage water. The water escaped through rock fissures underlying the volcanic surface rock on the reservoir floor.

Structure Number 11113　　Constructed - 1915
Stark Street　　Reinforced Concrete Deck Arch
Gresham vicinity, Multnomah County　　Ownership - Multnomah County

STARK STREET VIADUCT

This structure is a 68-foot filled spandrel, rib-type, reinforced concrete deck arch. The structure spans a small drainage that flows into the nearby Sandy River. On the Sandy River side of the bridge is an ornate overlook area with benches. Architectural treatment of the viaduct includes solid concrete railing with panels, stone masonry approach railing, and two ornamental lampposts. The lampposts are adorned with Ionic order column caps and pine cone replicas. The lamps are no longer present. The Stark Street Viaduct achieves its main distinction through this ornate treatment and its relationship to the original Columbia River Highway. The structure was built at the same period as the Columbia River Highway in the Columbia Gorge and has design elements similar to the structures on that highway. Stark Street served as one of the main routes connecting Portland with the scenic Columbia River Highway.

Structure Number 2379
Jacksonville Highway 272, M.P. 13.12
Provolt vicinity, Josephine County

Constructed - 1917
Reinforced Concrete Deck Arch
Ownership - State of Oregon

WILLIAMS CREEK BRIDGE

The Williams Creek Bridge is an 80-foot filled-spandrel, rib-type, concrete deck arch with a low rise. The bridge railings are a simple frame design and appear to be a later addition. The structure has a unique feature, the roadway rises to follow the shape of the arch. The bridge was constructed by Albert Anderson, a local contractor for Josephine County, and was acquired by the state in 1933.

Structure Number 332A
Sam's Valley Highway 271, M.P. 0.09
Gold Hill vicinity, Jackson County

Constructed - 1920
Reinforced Concrete Deck Arch
Ownership - State of Oregon

ROGUE RIVER (ROCK POINT) BRIDGE

This structure is an early major arch bridge completed under the design auspices of Conde B. McCullough, Oregon's noted bridge engineer. McCullough designed the Rock Point Bridge, originally on the old Pacific Highway, to conform with and complement the local landscape. The structure is 505 feet in length, and the main span is a single 113-foot open-spandrel, rib-type reinforced concrete deck arch. The structure has curved arch fascia curtain walls. The railing on the original main span is an urn-shaped balustrade with a band of dentils below. The ends of the main piers are bushhammered for textured contrast. This structure replaced a timber Howe truss covered bridge. The bridge was constructed by Parker and Banfield, Portland. Because of the great depth of the water at the bridge location, it was impossible to build falsework under the main arch span. The contractor solved the problem by building a temporary wood truss span over the river to give support to the forms.

Structure Number 25B14　　Constructed - 1922
N.W. Alexandra Avenue　　Reinforced Concrete Deck Arch
Portland, Multnomah County　　Ownership - City of Portland

N.W. ALEXANDRA AVENUE VIADUCT

Designed by Fred T. Fowler under the direction of Portland City Engineer Olaf Laurgaard, this 150-foot rib-type reinforced concrete deck arch is an important example of braced-spandrel construction in the tradition of the Latourell Creek Bridge (1914) on the Columbia River Highway. The arch ribs are not a continuous structure, but consist of ten separate chord members arranged in an arch form. Concrete trestle and beam approach spans are located at each end of the arch. The steel bar handrails replaced an original concrete precast panel railing. The culmination of Fred Fowler's bridge design career with the City of Portland was the S.W. Vista Avenue Viaduct, completed in 1926. (See Other Historic Bridges group.)

Structure Number 490A
Myrtle Creek Highway 237, M.P. 0.40
Myrtle Creek, Douglas County

Constructed - 1922
Reinforced Concrete Deck Arch
Ownership - State of Oregon

SOUTH UMPQUA RIVER (MYRTLE CREEK) BRIDGE

Six reinforced concrete deck girder spans and three open-spandrel rib-type concrete deck arch spans make up this 547-foot structure designed by Conde B. McCullough. As McCullough's first multi-span arch, the Myrtle Creek Bridge represents a design theme to which McCullough returned many times in his career. Typical of his attention to architectural details, McCullough provided this structure with arched fascia curtain walls, bracketing, and decorative sidewalk railing. This bridge replaced a four-span timber Howe truss bridge. The arch bridge was built on the old Pacific Highway by Lindstrom and Feigenson, Portland. The deck on this structure was widened in 1986.

Structure Number 357
Oswego Highway 3, M.P. 11.43
Oregon City-West Linn, Clackamas County

Constructed - 1922
Steel Half-Through Arch
Ownership - State of Oregon

WILLAMETTE RIVER (OREGON CITY) BRIDGE

The Willamette River Bridge at Oregon City is a 745-foot structure consisting of a 360-foot steel through arch and eleven concrete deck girder approach spans. The steel arch span is protected from corrosion by encasement in sprayed-on concrete (Gunite), which gives it the appearance of a concrete structure. The detailing--pylons, ornate bridge railing, arched fascia curtain walls, fluted Art-Deco main piers, and the use of bush-hammered inset panels--identify this structure as a Conde B. McCullough design and contributes to the significance of the structure. The Oregon City Bridge is one of only four half-through arch designs in the state. The bridge replaced a suspension span built at this site in 1888. The suspension cables were used to support the arch sections during construction of the current bridge. The bridge was built by A. Guthrie and Company, Portland, and was originally on the old Pacific Highway.

Structure Number 7C11 Constructed - 1924
West Broadway Street Reinforced Concrete Deck Arch
Seaside, Clatsop County Ownership - City of Seaside

NECANICUM RIVER (SEASIDE) BRIDGE

This structure is the only known patented Luten arch in Oregon. The design and construction methods devised by Daniel B. Luten, a consulting engineer from Indianapolis, were widely used nationwide between 1910 and 1930. This ornate concrete deck arch consists of three 41-foot filled-spandrel, barrel-type sections with integral approaches. The bridge contains decorative urn-shaped balustrade railings and lampposts. A southside monument pedestal contains a bronze bas-relief, "Old Oregon Trail," by noted Oregon sculptor Avard Fairbanks. (The companion north pedestal relief is missing.)

Structure Number 839
Oakland-Shady Highway 234, M.P. 12.21
Winchester, Douglas County

Constructed - 1924
Reinforced Concrete Deck Arch
Ownership - State of Oregon

NORTH UMPQUA RIVER (WINCHESTER) BRIDGE

Another of Conde B. McCullough's major arch bridges, the Winchester Bridge is a remarkable 884-foot structure consisting of seven 112-foot open spandrel, rib-type reinforced concrete deck arch spans and five concrete deck girder approach spans. A gothic arch motif is evident in the spandrel fascia curtain walls, the railing balustrade, and in the recessed panels above the end piers. Curved, decorative bracketing, observation balconies and a band of dentils add to the architectural interest of the structure. The bridge was dedicated to Robert A. Booth, one of the early State Highway Commission members. The structure replaced a three-span steel through truss bridge. The bridge was originally on the old Pacific Highway and was constructed by H. E. Doering, Portland.

Structure Number 1095
Boyd County Road
Boyd vicinity, Wasco County

Constructed - 1925
Reinforced Concrete Deck Arch
Ownership - Wasco County

FIFTEENMILE CREEK (ADKISSON) BRIDGE

The Fifteenmile Creek Bridge was designed by State Highway Bridge Engineer Conde B. McCullough, and built by George F. Reeves of Portland. A photograph of the original nameplate on the bridge identified the structure as the Adkisson Bridge. The structure, designed for Wasco County, is a single span reinforced concrete deck arch 148 feet in length. The main span is 120-feet long. Like many open-spandrel McCullough spans, the spandrel columns display curved arched fascia curtain walls, precast concrete railings with small arched openings, capped posts, ornate brackets supporting the railing, and a band of dentils below the railing. The tasteful architectural treatment of the structure and its exceptional pastoral setting contribute to the significance of this bridge.

Structure Number 600
The Dalles-California Highway 4, M.P. 141.68
Terrebonne vicinity, Jefferson County

Constructed - 1926
Steel Deck Arch
Ownership - State of Oregon

CROOKED RIVER (HIGH) BRIDGE

This highway span over the Crooked River Gorge provides one of the leading points of scenic grandeur in the central portion of the state. Designed by Conde B. McCullough, this bridge is 464 feet long and consists of a 330-foot two-hinged steel braced-spandrel deck arch. The structure was one of the highest bridges in the United States (at 295 feet from deck to streambed) when it was constructed. Architectural features include an ornamental concrete bridge railing and entrance pylons. A high line cableway was used in the erection of the structure. It is located just upstream from the 1911 Oregon Trunk Railroad steel arch bridge designed by Ralph Modjeski.

Structure Number 2459
Oregon Coast Highway 9, M.P. 127.61
Depoe Bay, Lincoln County

Constructed - 1927
Reinforced Concrete Deck Arch
Ownership - State of Oregon

DEPOE BAY BRIDGE

The Kuckenberg-Wittman Company of Portland built this single span reinforced concrete deck arch at the mouth of Depoe Bay in 1927. Designed by Conde B. McCullough, the structure maintains the beauty of the coastline and conforms with its environment. The Depoe Bay Bridge is an important example of McCullough's engineering and artistic treatment and a familiar Oregon Coast landmark. A stairwell on the bay side and a walkway at the north end of the bridge provide access to Depoe Bay State Park. The bridge is 312 feet in length, and the main span is a 150-foot rib deck arch with an open spandrel. The rib arches are three-hinged with skewback sections, and the spandrel fascia curtain walls are curved. The capacity of the bridge was increased in 1940, when an addition was made on the seaward side of the bridge.

Structure Number 1089
Otter Crest Loop Road
Otter Crest vicinity, Lincoln County

Constructed - 1927
Reinforced Concrete Deck Arch
Ownership - State of Oregon

ROCKY CREEK (BEN JONES) BRIDGE

The Rocky Creek Bridge, constructed on the Oregon Coast Highway in 1927, is one of the most beautifully sited bridges in the state. The bridge spans a small gorge in a picturesque section of the Oregon Coast. It is now bypassed and provides only local service. This structure is also known as the Ben Jones Bridge and is named for the "Father of the Coast Highway." This span is 360 feet long with a 160-foot-long reinforced concrete rib deck main arch. The structure has ten reinforced deck girder approach spans, five on either side of the main span. The open spandrel arch and approach columns have semicircular fascia curtain walls. It was designed by Conde B. McCullough and built by H.E. Doering, Portland.

Structure Number 576
Sam's Valley Highway 271, M.P. 2.65
Gold Hill, Jackson County

Constructed - 1927
Reinforced Concrete Deck Arch
Ownership - State of Oregon

ROGUE RIVER (GOLD HILL) BRIDGE

The Rogue River Bridge at Gold Hill was designed by Conde B. McCullough and is the only open spandrel barrel arch in the state. This particular design was chosen to provide high lateral strength for protection against high water. The main arch is 143 feet long and is constructed of reinforced concrete. The spandrel columns have semi-circular arched fascia curtain walls, and low arched curtain walls connect the approach columns. The O. N. Pierce Company of Portland built the structure, located on the old Pacific Highway.

Structure Number 3545-000-1.2
Forest Road 3545
Mount Hood vicinity, Hood River County

Constructed - 1928
Reinforced Concrete Deck Arch
Ownership - Mount Hood National Forest, United States Forest Service

SOUTH FORK HOOD RIVER (SAHALIE FALLS) BRIDGE

The Sahalie Falls Bridge was designed and constructed by the United States Bureau of Public Roads, the antecedent of the present Federal Highway Administration. The main span is an 82-foot open-spandrel rib-type reinforced concrete deck arch. A pedestrian area with a stone masonry water fountain at the east end of the bridge provides a viewpoint of Sahalie Falls. The bridge is on a bypassed segment of the Mt. Hood Loop Highway near the Meadows Ski Area. The original spelling of Sahalie was "Sahale." However, Sahalie is used on current USGS and State of Oregon maps.

Structure Number 1418
Redwood Highway 25, M.P. 0.14
Grants Pass, Josephine County

Constructed - 1931
Reinforced Concrete Half-Through Arch
Ownership - State of Oregon

ROGUE RIVER (CAVEMAN) BRIDGE

Three 150-foot reinforced concrete half-through arch spans and two 50-foot concrete deck girder approach spans make up this 550-foot structure across the Rogue River in Grants Pass. The bridge is one of only four half-through arch bridges in the state. Designed by Conde B. McCullough, the bridge is detailed with an ornate bridge railing, pylon lampposts, bracketing, and segmenting of the arch ribs. This is the fourth bridge to be located at this site, beginning with a timber bridge in 1886. The Caveman Bridge was built by J. K. Holt, Salem, and was originally on the old Pacific Highway. The name Caveman results from Grants Pass's location at the gateway to the Oregon Caves National Monument. The caveman serves as the civic symbol for Grants Pass and was selected as the name for the structure.

Structure Number 1172
Oregon Coast Highway 9, M.P. 327.64
Wedderburn-Gold Beach, Curry County

Constructed - 1931
Reinforced Concrete Deck Arch
Ownership - State of Oregon

ROGUE RIVER (GOLD BEACH) BRIDGE

Designated as a National Historic Civil Engineering Landmark by the American Society of Civil Engineers in 1982, this 1,898-foot structure was designed by Conde B. McCullough and dedicated to Isaac Lee Patterson, Oregon Governor from 1927 to 1929. One of the most notable bridges in the Pacific Northwest, the Gold Beach Bridge consists of seven 230-foot open-spandrel rib type reinforced concrete deck arch spans and eighteen concrete deck girder approach spans. Architectural and decorative features include fluted Art-Deco entrance pylons, dentils, ornate bracketing and sidewalk railing, fluted spandrel columns and detailed arched fascia curtain walls. When completed, the Rogue River bridge was the largest structure constructed by the State Highway Department. It was the first structure in America to be constructed with the Freyssinet method of decentering and stress control, named after the French inventor. The success of the bridge led to the wide use of prestressing techniques in concrete construction. Although completed in late 1931, the dedication of the bridge did not occur until May 28, 1932.

Structure Number 1499
Oregon Coast Highway 9, M.P. 64.73
Tillamook vicinity, Tillamook County

Constructed - 1931
Reinforced Concrete Through Tied Arch
Ownership - State of Oregon

WILSON RIVER BRIDGE

This bridge by Conde B. McCullough was the first concrete tied-arch span constructed in America. The main span of this 180-foot structure is a 120-foot reinforced concrete through tied arch (or bow-string arch) of elliptical shape. The choice of the tied arch was dictated by the location, where the cost of making large foundations to sustain the thrust of conventional arches would have been prohibitive. Ornate curved bracketing and ornamental sidewalk railing add detail to the bridge. The structure was built by the Clackamas Construction Company and replaced a steel through truss span at this site.

Structure Number 1113
Oregon Coast Highway 9, M.P. 178.35
Heceta Head, Lane County

Constructed - 1932
Reinforced Concrete Deck Arch
Ownership - State of Oregon

CAPE CREEK BRIDGE

The bridge over Cape Creek, one of Conde B. McCullough's most unique and attractive arch designs, abuts the Cape Creek Tunnel on the Oregon Coast Highway at Devil's Elbow State Park. The numerous columns and arches of the bridge, complemented by the elevation, create an image reminiscent of the Roman aqueducts, particularly the Pont du Gard, near Nimes, France. The double-tiered structure and the sheer number of arches promote a sense of rhythm that combines with the decorative bracketing and railing to make this a very handsome structure. The main span of the 619-foot structure is a 220-foot open-spandrel rib-type reinforced concrete deck arch. The Cape Creek Bridge was constructed by John K. Holt and the Clackamas Construction Company.

Structure Number 1617
Pacific Highway East 1E, M.P. 11.20
Gladstone, Clackamas County

Constructed - 1933
Steel Through Tied Arch
Ownership - State of Oregon

CLACKAMAS RIVER (MCLOUGHLIN) BRIDGE

Designed by Conde B. McCullough, this 720-foot, three-span steel through tied-arch structure was the recipient of the 1933 Annual Award of Merit from the American Institute of Steel Construction as the "Most Beautiful Steel Bridge, Class C." Decorative features of the bridge include an ornamental concrete bridge railing, gothic arch shaped openings in the main piers, and fluted, Art-Deco entrance pylons. The bridge was built by Lindstrom and Feigenson, Portland, in conjunction with the McLoughlin Boulevard project between Oregon City and Portland. The bridge was dedicated to the memory of Dr. John McLoughlin, a leading figure in the early development of the Pacific Northwest.

Structure Number 1582
Jefferson Highway 164, M.P. 6.24
Jefferson, Marion-Linn counties

Constructed - 1933
Reinforced Concrete Through Arch
Ownership - State of Oregon

SANTIAM RIVER (JEFFERSON) BRIDGE

Dedicated to early Oregon pioneer Jacob Conser, who operated a ferry at this site, this 780-foot structure consists of three 200-foot reinforced concrete through arch spans and four concrete deck girder approach spans. The ornate bridge railings, soffit bracketing, and extensively fluted and decorated entrance obelisks add to the beauty of the bridge and define it as a Conde B. McCullough design.

Structure Number 1820
Oregon Coast Highway 9, M.P. 141.68
Newport, Lincoln County

Constructed - 1936
Steel Half-Through Arch
Ownership - State of Oregon

YAQUINA BAY (NEWPORT) BRIDGE

The Yaquina Bay Bridge is an outstanding design and one of the most familiar and attractive bridges in Oregon. Constructed as one of five major bridges completed in 1936 under Oregon's Coast Bridges project, this Conde B. McCullough-designed structure is a combination of both steel and concrete arches. The main span of the 3,223-foot structure is a 600-foot steel through arch flanked by two 350-foot steel deck arches. There are five reinforced concrete deck arch secondary spans on the south end of the steel arches and fifteen concrete deck girder approach spans. Decorative elements include ornamental spandrel deck railing brackets, fluted entrance pylons, and a pedestrian plaza with elaborate stairways leading to observation areas. It was built by the Gilpin and General Construction companies and cost $1.3 million.

SUSPENSION BRIDGES

The suspension bridge is perhaps the most impressive of the bridge types. Usually large in size, with thin parabolic cables stretched between tall slender towers, the large suspension bridges in America are popular structural forms. They provide a fine, yet powerful, visual appearance and have become public monuments to bridge building. The Brooklyn Bridge and the Golden Gate Bridge, probably the best known bridges in America, are both suspension bridges.

Large suspension highway structures are expensive and are usually most appropriate for long spans where conditions such as deep water prohibit closely-spaced piers. Only three public highway suspension bridges are known in Oregon, the St. John's Bridge (1931) across the Willamette River in Portland, Multnomah County (shown on the following page); and the Crooked River Bridge (1962) on Lake Billy Chinook in Jefferson County (appendix F); the third crosses the Deschutes River nearby, and was built in 1962 in conjunction with the same reservoir project that constructed the Crooked River Bridge.

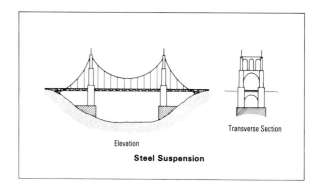

Steel Suspension

The St. John's Bridge is one of the most important bridges in the state because of its visual beauty, the fame of its designer, and the use of several technical innovations.

THE SUSPENSION BRIDGES

Bridge Name	Bridge Number	County	Type	Date
Willamette River (St. John's)	6497	Multnomah	Steel Suspension	1931

Structure Number 6497
Northeast Portland Highway 123, M.P. 0.91
Portland, Multnomah County

Constructed - 1931
Steel Suspension
Ownership - State of Oregon

WILLAMETTE RIVER (ST. JOHN'S) BRIDGE

Designed by internationally-famous bridge engineer David B. Steinman, the St. John's Bridge is one of the most significant structures in the state and the only large early steel suspension bridge. The 1,207-foot main span was the longest of its type in the world for many years. Steinman utilized many innovative features in the bridge, such as the highest concrete rigid frame pier in the world, the first use of main steel towers (400 feet high) without conventional diagonal bracing, and the use of prestressed rope strands (spun by John A. Roebling's Sons Company, of Brooklyn Bridge fame) instead of the conventional parallel wire cable construction. The bridge was built by Multnomah County and later acquired by the state. It is the tallest of the bridges in Portland with a 205-foot navigational clearance at zero water level. When the St. John's Bridge opened, the last of the Willamette River ferries in Portland was retired, ending a method of river crossing that lasted 83 years. St. John's Bridge is a designated Portland Historical Landmark.

MOVEABLE BRIDGES

The moveable bridge is utilized where the roadway would otherwise obstruct a navigable waterway and impede river traffic. Due to the large number of navigable waterways in the state, the moveable span bridge has played an important role in the development of Oregon's transportation system.

The three major types of moveable bridges, illustrated at right, are the swing, vertical lift, and bascule. The swing span represents the oldest of these three basic moveable bridge types. Swing spans were slow to operate, however, and were displaced by more efficient and quicker opening bascule and vertical-lift bridges. No swing span bridges have been constructed in Oregon since 1939.

Twenty-eight moveable span bridges remain in Oregon, 18 of which were constructed before 1941. Two of Oregon's moveable span bridges, the Columbia River (Interstate Northbound) Bridge (1917) and the Steel Bridge (1912) across the Willamette River in Portland, are already listed or eligible for the National Register. Six additional moveable structures have been identified as historically significant and are shown on the following pages.

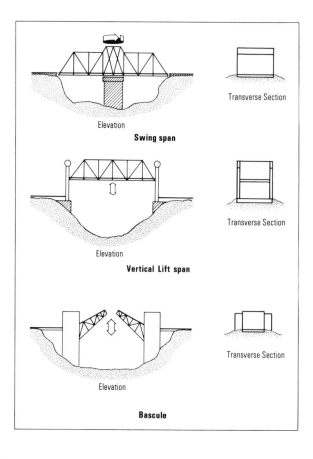

THE MOVEABLE BRIDGES

Bridge Name	Bridge Number	County	Type	Date
Willamette River (Hawthorne)	2757	Multnomah	Steel Through Truss (Parker) Vertical Lift	1910
Willamette River (Broadway)	6757	Multnomah	Steel Through Truss (Pennsylvania-Petit) Double-Leaf Bascule	1913
Coquille River	598	Coos	Steel Through Truss (Parker) Swing	1922
Willamette River (Burnside)	511	Multnomah	Steel Double-Leaf Bascule	1926
Siuslaw River (Florence)	1821E	Lane	Steel Double-Leaf Bascule	1936
Umpqua River (Reedsport)	1822	Douglas	Steel Through Truss (Parker) Swing	1936

Structure Number 2757
Madison Street-Hawthorne Boulevard
Portland, Multnomah County

Constructed - 1910
Steel Through Truss (Parker) Vertical Lift
Ownership - Multnomah County

WILLAMETTE RIVER (HAWTHORNE) BRIDGE

The Hawthorne Bridge is the oldest remaining highway structure across the Willamette River. The main span is a 244-foot steel through truss (Parker) vertical lift span, capable of a vertical movement of 110 feet and providing a lateral waterway clearance of 230 feet. Two electric motors lift the vertical deck lift span. The two towers are 165 feet tall. The bridge includes five steel through truss (Parker) secondary spans, each 220 feet in length, and thirteen concrete approach spans. The Hawthorne Bridge is the lowest of the Willamette River bridges in Portland, with 53 feet of clearance at low water, and consequently is raised more than any of the other drawbridges. This structure replaced a timber drawspan structure (Madison Street Bridge) built in 1891 and destroyed by fire in 1902. The Hawthorne Bridge has little architectural or decorative treatment. It was designed by Waddell and Harrington, Kansas City, and constructed by the Pennsylvania Steel Company, Portland, for a total cost of $511,000.

Structure Number 6757
Broadway Street
Portland, Multnomah County

Constructed - 1913
Steel Through Truss (Pennsylvania-Petit)
Double-Leaf Bascule
Ownership - Multnomah County

WILLAMETTE RIVER (BROADWAY) BRIDGE

The Broadway Bridge, designed by the internationally famous bridge designer Ralph Modjeski, is cited as "an important example of the Rall-type bascule span" by David Plowden in **Bridges: The Spans of North America** *(1974). The rarity and uniqueness of the Rall bascule structure add considerable technological interest to this structure. Built over a period of two years by the Pennsylvania Steel Company at a cost of $1.6 million, the bridge was the longest double-leaf bascule drawbridge in the world when constructed. The central span is a 297-foot steel through truss double-leaf bascule drawspan, providing 250 feet of lateral waterway clearance. The five secondary spans, four Pennsylvania-Petit steel through trusses and one Pratt steel through truss total 1,736 feet in length. An ornate vintage wrought iron bridge railing adjoins the sidewalks.*

Structure Number 598
Coquille-Bandon Highway 244, M.P. 17.08
Coquille, Coos County

Constructed - 1922
Steel Through Truss (Parker) Swing
Ownership - State of Oregon

COQUILLE RIVER BRIDGE

Designed under the auspices of Conde B. McCullough and constructed by local contractor A.B. Gidley, the Coquille River Bridge is a 235-foot steel through truss swing span. This structure is one of only six remaining swing spans on Oregon's highway system and is representative of this virtually obsolete moveable bridge type. (Four of Oregon's swing spans are in the final stages of bridge replacement planning.) The quicker and more efficient lifts and bascules have displaced the swing span system as the preferred moveable bridge types. The Coquille River Bridge is unique in its asymmetrical characteristics. Commonly referred to as a "bobtail" design, the swing span pivots on an off-center axis to obtain the maximum lateral waterway clearance. Once a busy commercial river site, the bridge is now rarely opened. A small riverside county park is near the structure.

Structure Number 511
Burnside Street
Portland, Multnomah County

Constructed - 1926
Steel Double-Leaf Bascule
Ownership - Multnomah County

WILLAMETTE RIVER (BURNSIDE) BRIDGE

The Burnside Bridge is a double-leaf bascule drawspan. It replaced the original 1894 wrought iron truss swing span structure. Two spans of the 1894 structure were moved to new locations and are the oldest highway bridges in Oregon (Bull Run River Bridge and the Sandy River Bridge on Lusted Road, both in Clackamas County). The Burnside Bridge has two 266-foot steel deck truss secondary spans and thirty-four steel deck girder approach spans for a total structure length of 2,308 feet. The bascule system for the bridge was designed by Joseph B. Strauss, who later designed San Francisco's Golden Gate Bridge. The principal engineer for the Burnside bridge was noted engineer Gustav Lindenthal. The original design concept is credited to I.G. Hendrick and Robert Kremers of Multnomah County, who were later replaced by Lindenthal. The Pacific Bridge Company constructed the bridge. Architectural treatment of the bridge includes an ornate spindle-type balustrade railing (wrought iron on the bascule section) and turreted operator shelters cantilevered from the massive main piers. The Burnside Bridge is distinguished as one of the most visually appealing of Portland's Willamette River bridges.

Structure Number 1821E
Oregon Coast Highway 9, M.P. 109.98
Florence, Lane County

Constructed - 1936
Steel Double-Leaf Bascule
Ownership - State of Oregon

SIUSLAW RIVER (FLORENCE) BRIDGE

This combination steel double-leaf bascule drawspan and reinforced concrete through tied arch structure is an outstanding example of the bridge engineering of Conde B. McCullough. Decorative treatment abounds with approach pylons, ornate railings, obelisk towers, and other details on the bridge. The Florence Bridge was one of five major structures built as part of Oregon's Coast Bridges project in 1936. Like the other bridges completed on the Oregon Coast Highway in 1936, the Florence Bridge replaced ferry service across one of Oregon's major rivers. The 140-foot central bascule drawspan is between two 154-foot concrete through tied (or bowstring) arch spans. Twenty concrete girder approach spans contribute to a total structure length of 1,568 feet. The bridge was built by the Mercer-Fraser Company, Eureka.

Structure Number 1822
Oregon Coast Highway 9, M.P. 211.11
Reedsport, Douglas County

Constructed - 1936
Steel Through Truss (Parker) Swing
Ownership - State of Oregon

UMPQUA RIVER (REEDSPORT) BRIDGE

One of the five major structures built during Oregon's Coast Bridges project, the Umpqua River Bridge is an important example of Conde B. McCullough's bridge engineering accomplishments. The steel through truss central span, at 430 feet, is the largest swing span structure in Oregon and is notable as a representative of this outdated moveable bridge technology. Only five other swing span bridges remain on Oregon's highway system. Two 154-foot reinforced concrete through tied (or bowstring type) arch spans are on either side of the central truss swing span. Though not as ornate as some of McCullough's bridges, the Reedsport Bridge has decorative railings, bracketing, and approach pylons. The bridge was built by Teufel and Carlson, Seattle.

SLAB, BEAM, AND GIRDER BRIDGES

The majority of the early bridges in Oregon are of the slab, beam, and girder type. A slab bridge consists of a single piece of reinforced concrete or other material supported by piers or abutments. The roadway and its support are integral in the slab bridge form. The beam and girder bridges may be of timber, steel or reinforced concrete. The deck of a beam structure is supported by transverse structural members framed into the main outer longitudinal structural members. In a girder bridge, the deck is supported by one or more longitudinal structural members without the transverse members.

This bridge type has not received much historic attention because of the normally common design, the relative uniformity of appearance, and the large numbers. The ODOT study, nonetheless, identified ten outstanding historic bridges of this type.

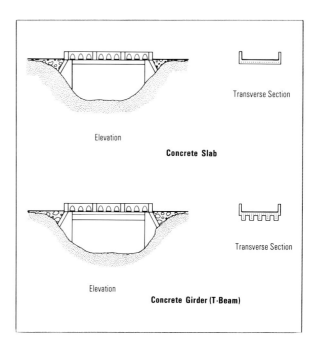

THE SLAB, BEAM, AND GIRDER HIGHWAY BRIDGES

Bridge Name	Bridge Number	County	Type	Date
Beaver Creek (Sandy River Overflow)	4522	Multnomah	Reinforced Concrete Deck Girder	1912
Dollarhide	3781	Jackson	Reinforced Concrete Deck Girder	1914
Old Mill Race	24T04	Umatilla	Reinforced Concrete Slab	1914
Steinman	3780	Jackson	Reinforced Concrete Deck Girder	1914
Fifteenmile Creek (Seufert)	308	Wasco	Reinforced Concrete Deck Girder	1920
Mill Creek (West Sixth Street)	464	Wasco	Reinforced Concrete Deck Girder	1920
Pringle Creek (Liberty Street S.E.)	1357	Marion	Reinforced Concrete Deck Girder	1928
Pringle Creek/Shelton Creek (Church Street S.E.)	608	Marion	Reinforced Concrete Deck Girder	1929
Chasm (Neahkahnie Mountain)	2723	Tillamook	Reinforced Concrete Deck Girder	1937
Necarney Creek	2311	Tillamook	Steel Deck Girder	1937

Structure Number 4522
Crown Point Highway 125, M.P. 1.74
Troutdale, Multnomah County

Construction - 1912
Reinforced Concrete Deck Girder
Ownership - State of Oregon

BEAVER CREEK (SANDY RIVER OVERFLOW) BRIDGE

This 40-foot concrete deck girder bridge was built in 1912 and is Oregon's oldest unaltered concrete girder bridge. The bridge has arched outer girder members and a heavy concrete railing typical of the era. A metal pipe outer railing appears to be a later addition. The bridge was built by Multnomah County and served on the old Columbia River Highway.

Structure Number 3781
Siskiyou Highway 273, M.P. 4.70
Steinman vicinity, Jackson County

Constructed - 1914
Reinforced Concrete Deck Girder
Ownership - State of Oregon

DOLLARHIDE OVERCROSSING

This 86-foot three-span concrete deck girder structure was built in 1914 for a cost of $5,900. A notable early example of the use of reinforced concrete in Oregon, it was constructed on the old Pacific Highway. The bridge's skewed design allows for both the crossing of the railroad and curvature of the roadway. Its heavy concrete railing was common during the 1910-1920 period. The structure was designed under the auspices of the state bridge engineer C.H. Purcell. J.W. Sweeney was the contractor.

Structure Number 24TO4
S.E. Court Place
Pendleton, Umatilla County

Constructed - 1914
Reinforced Concrete Slab
Ownership - City of Pendleton

OLD MILL RACE BRIDGE

This 28-foot structure is the oldest known concrete slab structure on Oregon's highway system. Designed by Pendleton city engineer Geary Kimbrell, it crosses a now filled mill race which served the original Pendleton Woolen Mills, northeast of the bridge. The concrete post and iron-pipe railing is unique in the state. The Pendleton Woolen Mills, nationally-famous for their quality woolen fabrics and garments, were established in 1893. The mill building in Pendleton was built in 1909, after C.P. Bishop purchased the company. At first, the primary operation at the Pendleton facility was producing blankets for sale to the Umatilla Indians, but the company expanded their operations and line of products over the years.

Structure Number 3780
Siskiyou Highway 273, M.P. 3.63
Steinman, Jackson County

Constructed - 1914
Reinforced Concrete Deck Girder
Ownership - State of Oregon

STEINMAN OVERCROSSING

Constructed on the old Pacific Highway in 1914, the Steinman Overcrossing is a 78-foot reinforced concrete deck girder. The railing design is similar to four structures on the Columbia River Highway, also constructed in 1914. This bridge is the only known switchback on Oregon's highway system where the roadway passes both under and over the structure. The Southern Pacific Railroad also passes under the structure. The overcrossing was constructed for the State Highway Department by J.W. Sweeney, Contractor. The bridge was designed under the auspices of state bridge engineer C.H. Purcell.

Structure Number 308
State (County) Road
The Dalles vicinity, Wasco County

Constructed - 1920
Reinforced Concrete Deck Girder
Ownership - Wasco County

FIFTEENMILE CREEK (SEUFERT) VIADUCT

This reinforced concrete girder structure derives its name Seufert Viaduct from two pioneer brothers who moved to Oregon in the early 1880s. The bridge was designed by Conde B. McCullough and constructed by the State Highway Department on the Columbia River Highway. The bridge was built under contract to the Colonial Building Company. The total length is 222 feet, consisting of one 22-foot span and five 40-foot spans. Parabolic arched fascia walls between the piers, ornate railing and brackets, and curved pier caps add architectural interest. The concrete fascia has bush-hammered inset panels for textural contrast. The structure is now on a county road (called State Road) and serves only light traffic. The old viaduct is near The Dalles Dam on the Columbia River.

Structure Number 464
West Sixth Street
The Dalles, Wasco County

Constructed - 1920
Reinforced Concrete Deck Girder
Ownership - City of The Dalles

MILL CREEK (WEST 6TH STREET) BRIDGE

One of the most ornate deck girder bridges in Oregon, the Mill Creek Bridge is a reinforced concrete structure designed by Conde B. McCullough. The 124-foot bridge was built in 1920 by Lindstrom and Feigenson, Contractors, for the City of The Dalles and is located on the route of the old Columbia River Highway. The ornate railing consists of urn-shaped balustrades with posts and concrete caps. Ten lampposts are spaced on the railing, but the original lanterns are missing. The bridge has arched fascia walls between the bridge piers. The fascia is bush-hammered for textural contrast with the smooth concrete.

Structure Number 1357
Pacific East Highway 1E, M.P. 50.59
Salem, Marion County

Constructed - 1928
Reinforced Concrete Deck Girder
Ownership - State of Oregon

PRINGLE CREEK (LIBERTY STREET, S.E.) BRIDGE

In the landscaped setting of Salem's Civic Center complex, this eight-span 326-foot concrete deck girder span was built during Salem's bridge construction program of 1928-29. During that period, about twelve new bridges were constructed in Salem, all designed by city bridge engineer R.A. Furrow. The structure features an ornamental precast bridge railing, keystone replicas atop the arched outer girder members, bush-hammered insets, ornate soffit bracketing, and a pedestrian walkway which passes under the structure adjacent to the creek. Liberty Street, S.E., serves northbound traffic. Southbound traffic uses the parallel Commercial Street, South, crossing of Pringle Creek, also completed in 1928.

Structure Number 608　　Constructed - 1929
Church Street, S.E.　　Reinforced Concrete Deck Girder
Salem, Marion County　　Ownership - City of Salem

PRINGLE CREEK/SHELTON CREEK (CHURCH STREET, S.E.) BRIDGE

Designed by Salem city bridge engineer R.A. Furrow, this reinforced concrete deck girder bridge is the most ornate of twelve or more city bridges built in 1928-29. The structure is over 500 feet long and spans both Pringle and Shelton creeks. The girder members of the bridge are arched and have bush-hammered inset panels. The piers are fluted, and curved brackets support the ornate railing and sidewalks. Vintage lampposts and lanterns are located at the ends of the bridge and on either side of a pedestrian stairway which leads down to Pringle Park. Along with the historic district in which it is located, this span was added to the Historic Register in October 1986.

Structure Number 2723
Oregon Coast Highway 9, M.P. 40.71
Manzanita vicinity, Tillamook County

Constructed - 1937
Reinforced Concrete Deck Girder
Ownership - State of Oregon

CHASM (NEAHKAHNIE MOUNTAIN) BRIDGE

Located on one of Oregon's most spectacular stretches of coastline, this 59-foot concrete deck girder represents a major engineering feat. Built on the almost vertical face of Neahkahnie Mountain, the curved outer girder member gives the structure the appearance of an arch. The stone masonry on the bridge is also used on the railings which extend nearly uninterrupted around the mountain. The impressive environmental setting, perched several hundred feet above the Pacific Ocean, can be enjoyed from several turn-out vista areas near the Chasm Bridge. Oswald West State Park borders the highway section. The structure was designed by state bridge engineer Glenn S. Paxson.

Structure Number 2311
Oregon Coast Highway 9, M.P. 39.53
Arch Cape vicinity, Tillamook County

Constructed - 1937
Steel Deck Girder
Ownership - State of Oregon

NECARNEY CREEK BRIDGE

This structure is a seven-span 602-foot steel deck girder supported on steel towers rising 90 feet above the creekbed. Attractively sited in Oswald West State Park, the bridge has a gothic arch balustrade railing and a broad, sweeping, horizontally-curved deck. This was one of the first steel girders with steel towers used in Oregon, and it heralded a period of emphasis on structural steel bridges which lasted until the 1960s. A bronze plaque notes its dedication to Samuel G. Reed, instrumental in the construction of the Neahkahnie Mountain section of the Oregon Coast Highway and a generous benefactor for Oswald West State Park. The structure was designed by state bridge engineer Glenn S. Paxson.

OLD COLUMBIA RIVER HIGHWAY BRIDGES

The old Columbia River Highway is the most historically significant twentieth century highway in Oregon. Built over a ten-year period from 1913 to 1922, the scenic highway was both environmentally sensitive to the magnificent Columbia River Gorge landscape (Figure 22) and ambitiously engineered. The highway was a major civic achievement and championed by such Oregon notables as Simon Benson, John B. Yeon, Samuel Hill, Julius Meier and Rufus Holman. Begun in the same year as the establishment of the State Highway Department (1913), the highway was developed under the auspices of that newly created organization in cooperation with the counties. The Columbia River Highway became a primary component of the initial state highway system adopted in 1914.

The old Columbia River Highway is a showplace of early highway engineering. In addition to much of the roadway and its alignment, most of the major engineering features still exist, including over 25 bridges and viaducts, three tunnels, half-tunnels, long stretches of dry masonry retaining walls, rustic rubble parapets, and pedestrian overlooks. This cliff-face highway along one of America's major rivers was the dream of Samuel Hill, who envisioned a roadway similar to the historic roads of Europe but built to the most modern standards of the day. Samuel C. Lancaster was the engineer in charge of the highway, particularly in Multnomah County. Along with John B. Yeon, Roadmaster, and Herbert Nunn, Lancaster surveyed, planned and constructed the highway.

FIGURE 22. *The 80-mile gorge of the Columbia River east of Portland posed an obstacle to highway building. Constructed between 1913 and 1922, the scenic Columbia River Highway was a major engineering feat. Crown Point Vista House, an observatory and rest area on the old highway, was completed in 1918.*

The early bridges on the old highway in Multnomah and Hood River counties were designed under the direction of Sam Lancaster and Charles H. Purcell, Oregon's first state bridge engineer. The actual designs are attributed primarily to H. K. Billner and L. W. Metzger, designing engineers with the State Highway Department. As the highway was completed eastward into Wasco County, the bridge designs became those of Conde B. McCullough, who joined the department in 1919. The bridges on the Columbia River Highway comprise one of the finest extant collections of early twentieth century reinforced concrete structures in America. Frequently of the arch form, the bridges were designed with great care to blend with their settings and to be light, graceful, and strong. Some were major engineering feats and others innovative in reinforcement and design.

In December 1983, 55 miles of the old highway, along with its associated structures and other features, were listed on the National Register. Subsequently, the highway has also been recognized as a National Historic Civil Engineering Landmark by the American Society of Civil Engineers. The National Park Service, United States Department of the Interior, is currently considering the scenic Columbia River Highway for National Historic Landmark designation.

Twenty-six major highway bridges located in the Columbia River Highway Historic District are summarized and then highlighted on the following pages. The bridges are in chronological order by construction date, which tends to coincide with their locations on the old highway from west-to-east within the historic district boundaries. (The historic district also includes bridges and structures not listed or pictured herein, such as pedestrian bridges, highway culverts, small slab spans, tunnels, retaining walls and railings.)

THE MAJOR COLUMBIA RIVER HIGHWAY BRIDGES

Bridge Name	Bridge Number	County	Type	Date
Sandy River (Troutdale)	2019	Multnomah	Steel Through Truss (Pratt)	1912
Sandy River (Stark Street)	11112	Multnomah	Steel Through Truss (Parker)	1914
Crown Point	4524	Multnomah	Reinforced Concrete Slab	1914
Latourell Creek	4527	Multnomah	Reinforced Concrete Deck Arch	1914
Young Creek (Shepperd's Dell)	4528	Multnomah	Reinforced Concrete Deck Arch	1914
Bridal Veil Falls	823	Multnomah	Reinforced Concrete Through Girder	1914
Wahkeena Falls	4533	Multnomah	Reinforced Concrete Slab	1914
West Multnomah Falls	840	Multnomah	Reinforced Concrete Slab	1914
Multnomah Creek	4534	Multnomah	Reinforced Concrete Deck Arch	1914
East Multnomah Falls	841	Multnomah	Reinforced Concrete Slab	1914
Oneonta Gorge Creek (Old)	4542	Multnomah	Reinforced Concrete Slab	1914
Horsetail Falls	4543	Multnomah	Reinforced Concrete Slab	1914
Moffett Creek	2194	Multnomah	Reinforced Concrete Deck Arch	1915
Tanner Creek	2062	Multnomah	Reinforced Concrete Deck Girder	1915
Toothruck and Eagle Creek	—	Multnomah	Reinforced Concrete Deck Girder	1915
Eagle Creek	—	Multnomah	Reinforced Concrete Deck Arch	1915
Ruckel Creek	—	Hood River	Reinforced Concrete Slab	1917
Gorton Creek	27C35	Hood River	Reinforced Concrete Slab	1918
Ruthton Point	273	Hood River	Reinforced Concrete Deck Girder	1918
Rock Creek	65C63	Wasco	Reinforced Concrete Slab	1918
Rock Slide	504	Hood River	Reinforced Concrete Deck Girder	1920
Mosier Creek	498	Wasco	Reinforced Concrete Deck Arch	1920
Hog Creek Canyon (Rowena Dell)	523	Wasco	Reinforced Concrete Slab	1920
Chenoweth Creek	506	Wasco	Reinforced Concrete Slab	1920
Dry Canyon Creek	524	Wasco	Reinforced Concrete Deck Arch	1921
Oneonta Gorge Creek (New)	7108A	Multnomah	Reinforced Concrete Deck Girder	1948

Structure Number 2019
Crown Point Highway 125, M.P. 1.85
Troutdale, Multnomah County

Constructed - 1912
Steel Through Truss (Pratt)
Ownership - State of Oregon

SANDY RIVER (TROUTDALE) BRIDGE

This structure and the Stark Street crossing of the Sandy River form the traditional gateways from Portland to the Columbia River Gorge and the scenic Columbia River Highway. Originally constructed by Multnomah County, the two-span Sandy River Bridge at Troutdale is now in state ownership and is the oldest metal truss bridge in state ownership. The two identical Pratt trusses with riveted connections are 162 feet each. The bridge was designed by Waddell and Harrington, Consulting Engineers, Kansas City, and erected by the Oregon Bridge and Construction Company, Portland. A timber covered bridge spanned the Sandy River prior to the construction of the present bridge.

Structure Number 11112
Stark Street Intersection with Crown Point Highway 125, M.P. 4.38
Gresham vicinity, Multnomah County

Constructed - 1914
Steel Through Truss (Parker)
Ownership - Multnomah County

SANDY RIVER (STARK STREET) BRIDGE
(Also, Auto Club Bridge)

Constructed in 1914, the Stark Street Bridge was one of the first truss bridges constructed by the newly formed (1913) State Highway Department. Charles H. Purcell was the first state bridge engineer. The structure was built by George H. Griffin, Portland Bridge Company, at a cost of $21,000. The two-span structure consists of a 200-foot Parker truss main span and a 77-foot Warren pony truss secondary span. The truss connections are riveted. The construction of the bridge was expedited because the timber truss at this location collapsed ironically on Good Roads' Day, April 25, 1914. The Portland Automobile Club grounds were located near the bridge, and consequently, the bridge was also historically known as the Auto Club Bridge.

Structure Number 4524
Crown Point Highway 125, M.P. 11.44
Crown Point, Multnomah County

Constructed - 1914
Reinforced Concrete Slab
Ownership - State of Oregon

CROWN POINT VIADUCT

The Crown Point Viaduct was the first structure started on the Multnomah County portion of the Columbia River Highway. Samuel C. Lancaster was the supervising engineer for both Multnomah County and the State Highway Department. Lancaster located the highway to encircle Crown Point, a promontory rising vertically 625 feet about the river. (Crown Point was designated a National Natural Landmark in August 1971.) The "half-viaduct" prevented unnecessary excavation or fill to establish a roadbed on the point. The structure is 560 feet long and consists of twenty-eight 20-foot reinforced concrete slab spans. Vista House, an observatory and rest stop dedicated to early Oregon pioneers, was completed on Crown Point in 1918.

Structure Number 4527
Crown Point Highway 125, M.P. 13.76
Latourell, Multnomah County

Constructed - 1914
Reinforced Concrete Deck Arch
Ownership - State of Oregon

LATOURELL CREEK BRIDGE

This arch structure is a three-span reinforced concrete braced-spandrel deck arch, each span 80 feet long. The bridge is credited to K.R. Billner, designing engineer with the State Highway Department. Billner adapted the principles of French bridge expert Armand Considere, who developed advanced measures of concrete reinforcement. A characteristic of the structure is its lightness, necessitated by the poor foundation conditions. The structure was built by the Pacific Bridge Company, Portland, and is located to take advantage of the view of Latourell Falls (249 feet) south of the bridge. The bridge is located in Guy W. Talbot State Park.

Structure Number 4528
Crown Point Highway 125, M.P. 14.98
Latourell vicinity, Multnomah County

Constructed - 1914
Reinforced Concrete Deck Arch
Ownership - State of Oregon

YOUNG CREEK (SHEPPERD'S DELL) BRIDGE

This graceful reinforced concrete deck arch has a main arch span of 100 feet and consists of two parabolic arch ribs with open spandrels. Designed by K.R. Billner under the supervision of Samuel C. Lancaster, the structure was constructed by the Pacific Bridge Company, Portland, at a cost of $10,800. A stairwell and trail to the falls originates at the east end of the bridge. The structure is located in Shepperd's Dell State Park. Shepperd's Dell was donated as parkland by the owner, George Shepperd, a local farmer of modest means, in memory of his wife.

Structure Number 823
Crown Point Highway 125, M.P. 16.01
Bridal Veil vicinity, Multnomah County

Constructed - 1914
Reinforced Concrete Through Girder
Ownership - State of Oregon

BRIDAL VEIL FALLS BRIDGE

One of only two known early reinforced concrete through girders in Oregon, the Bridal Veil Falls Bridge has a total length of 100 feet. The through girder is 50 feet long. The diagonally-placed support piers are another unique feature, so set to allow greater horizontal clearance under the structure. The bridge was designed by K.R. Billner, under the auspices of Samuel C. Lancaster. The Pacific Bridge Company of Portland built the structure. Bridal Veil Falls is downstream from the structure. When constructed, the bridge also spanned lumber flumes associated with the Bridal Veil Lumber Company.

Structure Number 4533
Crown Point Highway 125, M.P. 19.17
Multnomah Falls vicinity, Multnomah County

Constructed - 1914
Reinforced Concrete Slab
Ownership - State of Oregon

WAHKEENA FALLS BRIDGE

This small reinforced concrete slab span is 18 feet long. The solid railing has bush-hammered panels for textural contrast, and the abutments are faced with dry masonry. It was designed by K.R. Billner and built by the Pacific Bridge Company of Portland. The small bridge is located near Wahkeena Falls (vertical drop 242 feet) and is in a Mount Hood National Forest recreation site.

Structure Number 840
Crown Point Highway 125, M.P. 19.50
Multnomah Falls vicinity, Multnomah County

Constructed - 1914
Reinforced Concrete Slab
Ownership - State of Oregon

WEST MULTNOMAH FALLS VIADUCT

Both east and west of Multnomah Falls are long "half-viaducts" or "side-hill viaducts," built to carry the roadway in the narrow area between the steep hillside and railroad. With the uphill side of the viaduct resting on the slope and the downhill side elevated on columns, these viaducts are a successful alternative to cutting into the unstable slopes to excavate a roadbed. The West Multnomah Falls Viaduct is 400 feet long and consists of twenty 20-foot slab spans. The viaduct was designed by K.R. Billner, under the supervision of S.C. Lancaster, and was built by the Pacific Bridge Company, Portland.

Structure Number 4534
Crown Point Highway 125, M.P. 19.72
Multnomah Falls, Multnomah County

Constructed - 1914
Reinforced Concrete Deck Arch
Ownership - State of Oregon

MULTNOMAH CREEK BRIDGE

The Multnomah Creek Bridge, near the 620-foot drop of Multnomah Falls, is a noteworthy short-span arch and is a significant component of the old Columbia River Highway. This reinforced concrete deck arch is 67 feet in length. The barrel arch has solid spandrel walls and is 40 feet in length. The bridge was designed by K.R. Billner under the supervision of C.H. Purcell, State Bridge Engineer, and S.C. Lancaster, Assistant State Highway Engineer. It was constructed by the Pacific Bridge Company of Portland. The Multnomah Falls Lodge (in photo background) was completed in 1925 by the City of Portland. In the Cascadia or National Park rustic style, the stone lodge was designed by prominent Portland architect A.E. Doyle.

Structure Number 841
Crown Point Highway 125, M.P. 20.03
Multnomah Falls vicinity, Multnomah County

Constructed - 1914
Reinforced Concrete Slab
Ownership - State of Oregon

EAST MULTNOMAH FALLS VIADUCT

The East Multnomah Falls Viaduct is the longest viaduct on the old highway and is 860 feet in length. Consisting of 86 10-foot reinforced concrete slab spans, the East Multnomah Falls Viaduct is similar to the West Multnomah Falls Viaduct. The light, ornate arched railing adds to the appearance of the viaduct. This railing design is used on only four structures on the old Columbia River Highway.

Structure Number 4542 *Constructed - 1914*
Crown Point Highway 125, M.P. 21.93 (Bypassed) *Reinforced Concrete Slab*
Oneonta, Multnomah County *Ownership - State of Oregon*

ONEONTA GORGE CREEK (OLD) BRIDGE

This reinforced concrete slab bridge was designed by K.R. Billner and constructed by the Pacific Bridge Company of Portland. It is 80 feet in length and consists of four 20-foot spans. A stairwell is located at the west end for access to the creek and trail. As constructed, the highway at the bridge passed into the Oneonta Tunnel. The tunnel was closed in the late 1940s, and a new bridge parallel to this bridge was built.

Structure Number 4543
Crown Point Highway 125, M.P. 22.21
Oneonta vicinity, Multnomah County

Constructed - 1914
Reinforced Concrete Slab
Ownership - State of Oregon

HORSETAIL FALLS BRIDGE

Very similar in design to the Oneonta Gorge Creek Bridge, the Horsetail Falls Bridge is a 60-foot reinforced concrete slab span type, consisting of three 20-foot spans. The ornate arch railing design is repeated on the Oneonta Gorge Creek Bridge, East Multnomah Falls Viaduct and West Multnomah Falls Viaduct. (The only other known use of this railing design in Oregon is the Steinman Overcrossing, Jackson County, also constructed in 1914.) The structure, built in 1914, is attributed to K.R. Billner, designing engineer at the State Highway Department, under the supervision of Samuel C. Lancaster. The contractor was the Pacific Bridge Company of Portland. (Horsetail Falls, located near the bridge, has a vertical drop of 221 feet.)

Structure Number 2194
Columbia River Highway 2 (Interstate 84),
M.P. 38.98 (Bypassed)
Bonneville vicinity, Multnomah County

Constructed - 1915
Reinforced Concrete Deck Arch
Ownership - State of Oregon

MOFFETT CREEK BRIDGE

This structure was one of the outstanding engineering features on the Columbia River Highway. The low-rise reinforced concrete deck arch has a clear span of 170 feet and rises only 17 feet in that distance. The overall length of the structure is 205 feet. When it was constructed in 1915, it was the longest three-hinged flat arch bridge in the United States. The three hinges are massive cast iron with 4-1/2-inch steel pins. The bridge is attributed to L.W. Metzger, a designer under the guidance of C.H. Purcell, the first state bridge engineer. The bridge was bypassed in 1950 and is now abandoned.

Structure Number 2062
Columbia River Highway 2 (Interstate 84),
M.P. 40.27 (Bypassed)
Bonneville vicinity, Multnomah County

Constructed - 1915
Reinforced Concrete Deck Girder
Ownership - State of Oregon
(Fish and Wildlife Commission)

TANNER CREEK BRIDGE

Bypassed and no longer in use, the Tanner Creek Bridge is a reinforced concrete deck girder, 60 feet in length. The bridge is located near the Interstate 84 entrance to the Bonneville Dam and is now owned by the Oregon Fish and Wildlife Commission. Completed in 1915, the bridge was constructed by the State Highway Department. Charles H. Purcell was the state bridge engineer, and Samuel Lancaster was the engineer for the Columbia River Highway.

Structure Numbers Unknown
Columbia River Highway 2 (Interstate 84),
M.P. 41.25 (Bypassed)
Bonneville vicinity, Multnomah County

Constructed - 1915
Reinforced Concrete Deck Girder
Ownership - State of Oregon

TOOTHROCK AND EAGLE CREEK VIADUCTS

These reinforced concrete deck girder viaducts are located on an abandoned portion of the old highway above the Toothrock Tunnel (1936), near Bonneville Dam. Although partially hidden by trees and other vegetation, the old viaducts are visible from the westbound lanes of Interstate 84. The viaducts curve around the mountainside and are about 224 feet long and 23 feet wide. Charles H. Purcell was the state bridge engineer. The viaducts were abandoned when the new alignment of the highway was completed (including the Toothrock Tunnel), in conjunction with the construction of the Bonneville Dam in the late 1930s.

Structure Number Unknown
Exit Road, Columbia River Highway 2
(Interstate 84), M.P. 41.55
Bonneville vicinity, Multnomah County

Constructed - 1915
Reinforced Concrete Deck Arch
Ownership - State of Oregon

EAGLE CREEK BRIDGE

The Eagle Creek Bridge is one of the most attractive structures on the original Columbia River Highway. The historic feeling is enhanced by the masonry facing and stone parapets. The reinforced concrete deck arch is 100 feet in length. The semi-circular arch has three rib arches and is 60 feet in length. At the ends of the arch span are 20-foot concrete slab spans. A pedestrian overlook is at the west end. The bridge was designed by K.R. Billner, Designing Engineer, State Highway Department. The bridge serves the eastbound exit from Interstate 84 to the state fish hatchery and recreation areas at Eagle Creek.

Structure Number Unknown
Forest Trail 405 from the Eagle Creek Campground
Cascade Locks vicinity, Hood River County

Constructed - 1917
Reinforced Concrete Slab
Ownership - Mount Hood National Forest,
United States Forest Service

RUCKEL CREEK BRIDGE

The Ruckel Creek Bridge is one of the smaller structures built on the scenic Columbia River Highway. A simple slab span, 10 feet in length, was all that was required to cross Ruckel Creek. The concrete abutments are faced with stone, and the railing is the arched masonry parapet commonly bordering the scenic highway in the Columbia Gorge. The bridge plans are signed by L.W. Metzger, State Highway Department. The bridge now serves a Mount Hood National Forest trail, accessible from the Eagle Creek Campground.

Structure Number 27C35
Wyeth County Road 605
Wyeth vicinity, Hood River County

Constructed - 1918
Reinforced Concrete Slab
Ownership - Hood River County

GORTON CREEK BRIDGE

The Gorton Creek Bridge is a 50-foot reinforced concrete slab structure, consisting of three slab spans (two 15-foot and one 20-foot spans). This simple bridge has a standard precast concrete railing with vertical openings, posts and caps. The bridge was designed by L.W. Metzger and was constructed by A.D. Kern for the State Highway Department. Now located on a county road, the bridge is visible from Interstate 84. A Mount Hood National Forest campground is east of the bridge. Gorton Creek was named for Edwin Gorton, who had a homestead along its banks in the 1890s.

Structure Number 273
Columbia River Highway 2 (Interstate 84),
M.P. 60 (approx.) (Bypassed)
Hood River vicinity, Hood River County

Constructed - 1918
Reinforced Concrete Deck Girder
Ownership - State of Oregon

RUTHTON POINT VIADUCT

This 50-foot half-viaduct is located on an abandoned section of the old highway west of Hood River at Ruthton Point. Near the westbound lanes of Interstate 84, the reinforced concrete deck girder structure consists of three spans--two 20-foot and one 10-foot in length. L.W. Metzger was the designing engineer for the State Highway Department.

Structure Number 65C63　　Constructed - 1918
Hood River (County) Road　　Reinforced Concrete Slab
Mosier, Wasco County　　Ownership - Wasco County

ROCK CREEK BRIDGE

The Rock Creek Bridge is a 44-foot reinforced concrete slab span structure, consists of two 22-foot slabs. The simple wood railing may not be original, but echoes a wood railing design used along long stretches of the highway in lieu of the rustic arched masonry parapets. The Rock Creek Bridge is now in county ownership, but was originally built by the State Highway Department. L. W. Metzger was the designing engineer.

Structure Number 504
Old Columbia River Highway Drive (County Road)
Hood River vicinity, Hood River County

Constructed - 1920
Reinforced Concrete Deck Girder
Ownership - Hood River County

ROCK SLIDE VIADUCT

Only 34 feet long, the Rock Slide Viaduct is a short structure not far from the Mosier Twin tunnels. Although visible from Interstate 84, the viaduct is not readily apparent when driving the old highway (now a Hood River County road). The uninterrupted roadway surface and the continuous arched rubble parapet railing across the viaduct make it difficult to identify the bridge span from the road. Noted Oregon State Bridge Engineer Conde B. McCullough had recently joined the Highway Department and signed the structure plans.

Structure Number 498
Mosier-The Dalles Highway 292, M.P. 0.64
Mosier, Wasco County

Constructed - 1920
Reinforced Concrete Deck Arch
Ownership - State of Oregon

MOSIER CREEK BRIDGE

A notable feature on the old Columbia River Highway, this reinforced concrete deck arch structure is 182 feet long and consists of a 110-foot rib arch and concrete slab approaches. It was designed by Conde B. McCullough, State Bridge Engineer. Lindstrom and Feigenson, Contractors, were the builders.

Structure Number 523
Mosier-The Dalles Highway 292, M.P. 5.97
Rowena vicinity, Wasco County

Constructed - 1920
Reinforced Concrete Slab
Ownership - State of Oregon

HOG CREEK CANYON (ROWENA DELL) BRIDGE

This small bridge is a 20-foot reinforced concrete slab span and was designed under the auspices of Conde B. McCullough, State Bridge Engineer. The railing design is unique to the structures in the historic district area. Rowena Dell, a canyon of basaltic rock between Rowena and Mosier, was originally known as Hog Creek Canyon. Local legend is that herds of hogs were grazed in the canyon to eradicate the rattlesnakes. After the highway was built, a more elegant name was desired, and Rowena Dell was selected.

Structure Number 506
Mosier-The Dalles Highway 292, M.P. 14.99
The Dalles vicinity, Wasco County

Constructed - 1920
Reinforced Concrete Slab
Ownership - State of Oregon

CHENOWETH CREEK BRIDGE

This 60-foot reinforced concrete slab span structure is the eastern boundary of the Columbia River Highway Historic District and lies near The Dalles city limits. The Chenoweth Creek Bridge consists of three 20-foot spans. The original railing has been replaced with metal guardrail. A.D. Kern of Portland constructed the bridge for the State Highway Department. Conde B. McCullough was the state bridge engineer. Chenoweth Creek was named for Justin Chenoweth, a prominent Oregon pioneer.

Structure Number 524
Mosier-The Dalles Highway 292, M.P. 6.64
Rowena vicinity, Wasco County

Constructed - 1921
Reinforced Concrete Deck Arch
Ownership - State of Oregon

DRY CANYON CREEK BRIDGE

The Dry Canyon Creek Bridge rivals Shepperd's Dell Bridge in Multnomah County for excellence in design and compatibility with its environmental setting. This small 75-foot reinforced concrete rib arch spans a deep ravine 600 feet above the Columbia River. The open spandrel arch structure was designed by Conde B. McCullough, State Bridge Engineer, and is one of his early arch designs. The bridge was built by state forces and supervised by resident bridge engineer Christ Fauerso.

Structure Number 7108A
Crown Point Highway 125, M.P. 21.93
Oneonta, Multnomah County

Constructed - 1948
Reinforced Concrete Deck Girder
Ownership - State of Oregon

ONEONTA GORGE CREEK (NEW) BRIDGE

The Oneonta Gorge Creek replacement bridge is a 48-foot reinforced concrete deck girder, completed in 1948. This bridge was constructed to replace the old bridge (Oneonta Gorge Creek Bridge, #4542), bypassed when the Oneonta Tunnel was closed. The revised highway alignment skirts the tunnel to the north and also required the relocation of the railroad tracks. The open steel bar railing adds a distinctive feature to this short bridge. The bridge was designed by Glenn S. Paxson, State Bridge Engineer.

COVERED BRIDGES

Oregon has the largest collection of covered bridges in the West and one of the largest in the nation. The covered bridge-building tradition in Oregon dates from the 1850s. Out of necessity, pioneers built with the materials at hand. Douglas fir was abundant in western Oregon and well-suited to bridge construction. Housings were placed over the timber truss and chords to protect these members from the damp climate of western Oregon, thus increasing their useful life. The heyday of covered bridge building in Oregon occurred between 1905 and 1925, when there were an estimated 450 covered bridges in the state.

By 1977, the number of covered bridges in the state had dwindled to 56 structures. Because of public concern for the decreasing number of covered bridges, the State Historic Preservation Office nominated the remaining covered bridges to the National Register. The nomination of the Oregon Covered Bridge Thematic Group was coordinated with the bridge owners--cities, counties and private parties. (Although once common on state routes, the last state-owned housed structure, Ritner Creek Bridge in Polk County, was relocated in 1976 and turned over to the county.)

At the express request of the city and county owners concerned, ten of Oregon's covered bridges were excluded from the thematic group nomination.

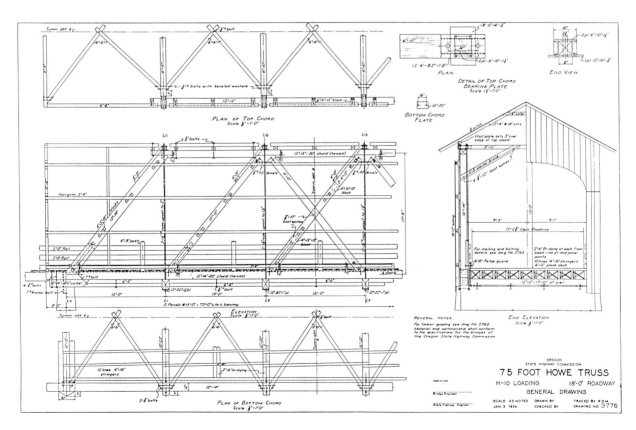

FIGURE 23. This 1934 standard plan for a 75-foot Howe truss covered bridge is one of several plans developed by the State Highway Department. These standard designs for covered bridges were not only used by the state, but were also available to local governments. Many of Oregon's surviving covered bridges are examples of standard state designs.

The excluded covered bridges were the following:

Name	County
North Fork Siuslaw River (Meadows)	Lane
Pass Creek (Drain)	Douglas
Calapooya Creek (Rochester)	Douglas
Crabtree Creek (Hoffman)	Linn
Thomas Creek (Jordan)	Linn
South Myrtle Creek (Neal Lane)	Douglas
Thomas Creek (Gilkey)	Linn
Little River (Cavitt Creek)	Douglas
Crabtree Creek (Bohemian Hall)	Linn
Thomas Creek (Shimanek)	Linn

In November 1979, 46 of Oregon's 56 covered bridges were subsequently listed on the National Register.

Since 1979, four covered bridges have been lost due to floods and displacements: *

Name	County
North Fork Siuslaw River (Meadows)	Lane
Siletz River (Sam's Creek)	Lincoln
Yaquina River (Elk City)	Lincoln
Mosby Creek (Brumbaugh)	Lane

Three of the lost covered bridges were listed on the National Register. The North Fork Siuslaw River (Meadows) Covered Bridge was not listed at the request of Lane County.

Currently, the population of covered bridges in Oregon numbers 52.† All of these covered bridges are illustrated in this document, except for the Coast Fork Willamette River (Chambers) Covered Bridge, Lane County, which was purposely not included. Although listed on the National Register, Chambers is a railroad bridge, a structure use not included in this historic highway bridges study.

The remaining 42 highway covered bridges listed on the National Register in 1979 are summarized below and shown on the following pages. The bridges are arranged chronologically by date of construction. (Appendices C and D are useful supplements to this section as they list the covered bridges by location and by structure type.)

The nine covered bridges excluded from the covered bridges nomination are shown in other places in this document. The ODOT historic bridges study evaluated five of the bridges for National Register eligibility--Pass Creek (Drain), Calapooya Creek (Rochester), Crabtree Creek (Hoffman), South Myrtle Creek (Neal Lane), and Thomas Creek (Gilkey). These five covered bridges are included in the Study Identified Bridges section. Because replacement was urgently needed, a request for determination of eligibility on the Thomas Creek (Jordan) Covered Bridge was specially prepared. That bridge was subsequently found National Register eligible and is included in the Other Historic Bridges section. The relatively recent construction dates of the Little River (Cavitt Creek), 1943; Crabtree Creek (Bohemian Hall), 1947; and Thomas Creek (Shimanek), 1966, covered bridges excluded them from the cut-off date of the historic bridges study. These bridges were, however, examined because of their special type and are shown with the Notable Post-1940 Construction Highway Bridges in Appendix F.

---

* As of November 1988, there are 49 standing covered bridges. Since 1985, Rickreall Creek Bridge (Pumping Station) collapsed into Rickreall Creek and three others—Weddle, Crabtree Creek (Bohemian Hall), and Horse Creek—are in storage for possible future use.

†This total does not include the Johnson Creek (Cedar Crossing) Bridge, built by Multnomah County on S.E. Deardorff Road in 1982. Although housed, this attractive contemporary structure contains no truss and is a timber (Glulam) stringer structure.

THE COVERED BRIDGES

Bridge Name	Bridge Number	County	Type	Date
Drift Creek	10803	Lincoln	Timber Howe Truss	1914
Rickreall Creek (Pumping Station) *	—	Polk	Timber Howe Truss	1916
Abiqua Creek (Gallon House)	5381	Marion	Timber Howe Truss	1917
Applegate River (McKee)	29C471	Jackson	Timber Howe Truss	1917
Alsea River (Hayden)	14538	Benton	Timber Howe Truss	1918
Five Rivers (Fisher School)	90101	Lincoln	Timber Howe Truss	1919
Lost Creek	29C262	Jackson	Timber Queenpost Truss	1919
Grave Creek	141005	Josephine	Timber Howe Truss	1920
Mosby Creek	39C241	Lane	Timber Howe Truss	1920
Lost Creek (Parvin)	19-1W-21	Lane	Timber Howe Truss	1921
Sandy Creek (Remote)	4037	Coos	Timber Howe Truss	1921
Antelope Creek	29C202	Jackson	Timber Queenpost Truss	1922
Coyote Creek (Battle Creek)	39C409	Lane	Timber Howe Truss	1922
Row River (Currin)	39C242	Lane	Timber Howe Truss	1925
Wildcat Creek	39C446	Lane	Timber Howe Truss	1925
Yaquina River (Chitwood)	41C09	Lincoln	Timber Howe Truss	1926
Evans Creek (Wimer)	29C211	Jackson	Timber Queenpost Truss	1927
Ritner Creek	1251	Polk	Timber Howe Truss	1927
Lake Creek (Nelson Mountain)	39C386	Lane	Timber Howe Truss	1928
Elk Creek (Roaring Camp)	—	Douglas	Timber Howe Truss	1929
Horse Creek *	16-5E-24	Lane	Timber Howe Truss	1930
Mosby Creek (Stewart)	39C243	Lane	Timber Howe Truss	1930
Calapooia River (Crawfordsville)	12819	Linn	Timber Howe Truss	1932
Deadwood Creek	16-9W-25	Lane	Timber Howe Truss	1932
Fall Creek (Unity)	14721	Lane	Timber Howe Truss	1936
Marys River (Harris)	1441	Benton	Timber Howe Truss	1936
Thomas Creek (Hannah)	12948	Linn	Timber Howe Truss	1936
Thomas Creek (Weddle) *	12935	Linn	Timber Howe Truss	1937
Fall Creek (Pengra)	18-1W-32	Lane	Timber Howe Truss	1938
McKenzie River (Goodpasture)	39C118	Lane	Timber Howe Truss	1938
Mill Creek (Wendling)	39C174	Lane	Timber Howe Truss	1938
Mohawk River (Earnest)	39C176	Lane	Timber Howe Truss	1938
North Fork Yachats River	12037	Lincoln	Timber Queenpost Truss	1938
Crabtree Creek (Larwood)	12876	Linn	Timber Howe Truss	1939
North Fork of the Middle Fork Willamette River (Office)	—	Lane	Timber Howe Truss	1944
Middle Fork Willamette River (Lowell)	19-1W-23	Lane	Timber Howe Truss	1945
South Fork Santiam River (Short)	14025	Linn	Timber Howe Truss	1945
Row River (Dorena)	21-2W-24A	Lane	Timber Howe Truss	1949
Willamette Slough (Irish Bend)	45005-17	Benton	Timber Howe Truss	1954
South Umpqua River (Milo Academy)	—	Douglas	Steel Through Girder	1962
Swalley Canal (Rock O' the Range)	—	Deschutes	Timber Deck Girder	1963
McKenzie River (Belknap)	39C123	Lane	Timber Howe Truss	1966

* See note on facing page.

Structure Number 10803
Drift Creek County Road
Lincoln City vicinity, Lincoln County

Constructed - 1914
Timber Through Truss (Howe) Covered Bridge
Ownership - Lincoln County

DRIFT CREEK BRIDGE

The Drift Creek Bridge is the oldest covered span in Oregon. It also has the distinction of being the covered bridge closest to the Oregon Coast, only 1-1/2 miles from the Pacific Ocean. Built by Lincoln County, the 66-foot housed Howe truss structure cost about $1,800 in 1914. The bridge is closed to vehicular traffic but serves pedestrian traffic. The old bridge has flared board-and-batten siding, arched portals, ribbon daylighting and wooden flooring. Lincoln County has maintained the bridge as a historical exhibit-in-place since 1965.

Structure Unnumbered
Private Road off Robb Mill
County Road 763*
Dallas vicinity, Polk County

Constructed - 1916
Timber Through Truss (Howe)
Covered Bridge
Ownership - Private

RICKREALL CREEK (PUMPING STATION) BRIDGE *

The plainest of Oregon's covered bridges is the Pumping Station Bridge near Dallas in Polk County. Of very early construction (1916), the 84-foot covered bridge includes an exposed Howe truss. (The sides of the bridge were originally covered, but the corrugated metal siding has been removed.) The bridge was built by the Dallas Water Company, a private company owned and operated by H.V. Gates. The narrow bridge was used for traffic crossing the creek that served the water works. The roof is covered with corrugated metal. The ends are open as well, and cantilevered buttresses (also called "sway braces") strengthen the sides of the structure.

* This structure collapsed into Rickreall Creek during highwater in November 1986.

Structure Number 5381
Gallon House County Road 647
Silverton vicinity, Marion County

Constructed - 1917
Timber Through Truss (Howe) Covered Bridge
Ownership - Marion County

ABIQUA CREEK (GALLON HOUSE) BRIDGE

The Gallon House Bridge is the last remaining covered bridge in Marion County and one of the oldest covered bridges in Oregon. This structure is an 84-foot housed Howe truss with a timber roadway deck. Gallon House has rectangular portals, narrow ribbon openings at the eaves, and board-and-batten siding. The bridge is open to traffic. Marion County completed major repairs to the bridge in 1985. According to local tradition, the bridge owes its name to an unauthorized liquor dispensary which operated at the north approach prior to prohibition. During that time, Silverton to the south was a dry town, while Mount Angel to the north was not.

Structure Number 29C471
Applegate County Road
Ruch vicinity, Jackson County

Constructed - 1917
Timber Through Truss (Howe) Covered Bridge
Ownership - Jackson County

APPLEGATE RIVER (McKEE) BRIDGE

The fourth oldest covered bridge in Oregon, the McKee Bridge was built in 1917 by contractor Jason Hartman of Jacksonville on land donated by stage station operator, Adelbert "Deb" McKee. The bridge was a rest stop at the half-way point between Jacksonville and the Blue Ledge Copper Mine. Relief horses were kept at this location for hauling ore in the early days of the bridge. The 122-foot housed Howe truss structure was closed to vehicular traffic in 1956. Local volunteers have since maintained the bridge. McKee Picnic Ground, a Rogue National Forest facility, is at the west end of the bridge along the Applegate River. The bridge contains truncated rectangular portal arches, ribbon openings at the eaves on both sides, five irregularly spaced windows on the south elevation, and exposed cantilevered buttresses.

Structure Number 14538　　Constructed - 1918
Hayden County Road　　Timber Through Truss (Howe) Covered Bridge
Alsea vicinity, Benton County　　Ownership - Benton County

ALSEA RIVER (HAYDEN) BRIDGE

The 91-foot Hayden Bridge is among the oldest covered bridges in the state. Seven covered bridges in Oregon remain that were constructed before 1920, including the Hayden Bridge. In addition, the Hayden Bridge is the oldest of Benton County's three covered bridges, all Howe trusses. Its original portal design has been changed to allow larger loads. Narrow ribbon openings under the side wall eaves allow some interior light, and the sides are flared.

Structure Number 90101 *Constructed - 1919*
Five Rivers County Road *Timber Through Truss (Howe) Covered Bridge*
Fisher, Lincoln County *Ownership - Lincoln County*

FIVE RIVERS (FISHER SCHOOL) BRIDGE

An early covered span, the Fisher School Bridge was constructed in 1919 and incorporates a 72-foot housed Howe truss. Lincoln County contracted with Otis Hamer to build the bridge for about $2,500. Design features include semi-elliptical portal arches, ribbon openings, flared side walls, wood piers and cribs. The bridge is now closed to vehicles, and a modern concrete bridge nearby handles daily traffic. Five Rivers is named for the multiplicity of creeks which feed this tributary of the Alsea River.

Structure Number 29C262
Lost Creek County Road
Lakecreek vicinity, Jackson County

Constructed - 1919
Timber Through Truss (Queenpost) Covered Bridge
Ownership - Jackson County

LOST CREEK BRIDGE

The Lost Creek Bridge at 39 feet is the shortest of the Oregon covered bridges and is also one of the oldest. The bridge has been bypassed and is closed to vehicular traffic. The housed queenpost truss is modified by cross members within the truss arrangement. The bridge does not have the usual portal arches and is open at the ends. Other architectural features include ribbon openings at the eaves and exposed cantilever buttresses.

Structure Number 141005
Sunny Valley-Placer County Road
Sunny Valley vicinity, Josephine County

Constructed - 1920
Timber Through Truss (Howe) Covered Bridge
Ownership - Josephine County

GRAVE CREEK BRIDGE

Because of its proximity to Interstate 5, the Grave Creek Bridge is Oregon's most viewed covered bridge. Originally built on the Pacific Highway (U.S. 99) in 1920, the structure was built by the State Highway Department and is a standardized state covered bridge plan. The housed Howe truss is 105 feet long. The original semi-elliptical portal arches have been reshaped to provide increased height clearance. Architectural features include false end heads. The Grave Creek Bridge is the only remaining covered bridge in Josephine County.

Structure Number 39C241
Layng County Road 2542
Cottage Grove vicinity, Lane County

Constructed - 1920
Timber Through Truss (Howe) Covered Bridge
Ownership - Lane County

MOSBY CREEK BRIDGE

Mosby Creek Bridge is the oldest in Lane County's collection of eighteen covered highway bridges, the largest of any county in Oregon. Built in 1920 at a cost of $4,125, the Mosby Creek Bridge is a housed Howe truss, 90 feet in length. Typical of the smaller bridges in Lane County, the bridge has semi-circular portal arches, ribbon openings at the roof line, and board-and-batten cladding.

Structure Number 19-1W-21 *Constructed - 1921*
Parvin County Road 6122 *Timber Through Truss (Howe) Covered Bridge*
Dexter vicinity, Lane County *Ownership - Lane County*

LOST CREEK (PARVIN) BRIDGE

The Parvin Covered Bridge was constructed by Lane County in 1921 at a cost of $3,600. George W. Breeding was the contractor. The structure is a 75-foot housed Howe truss and is a fine example in Oregon's covered bridge tradition. The current bridge replaced a span erected at this site in the late 1880s. The bridge was named for an early pioneer family. Distinctive features include the ribbon openings under the eaves and the truncated rectangular portal arches. The bridge was closed to vehicular traffic in 1974 and now serves as a footbridge for local residents.

Structure Number 4037
Sandy Creek County Road
Remote, Coos County

Constructed - 1921
Timber Through Truss (Howe) Covered Bridge
Ownership - Coos County

SANDY CREEK (REMOTE) BRIDGE

The 60-foot Remote Bridge is the only remaining covered span in Coos County. Bypassed in 1949, the Remote Bridge was formerly in state ownership on State Route 42. The truss consists of two crossed Howe truss members on each chord, a rarity in short covered trusses. The design of the bridge with large side openings is similar to those found in Linn County. Until recently, the bridge was in poor condition but has now been restored by volunteers. In September 1984, the bridge was dedicated as a Coos County park and serves pedestrian uses.

Structure Number 29C202
Antelope County Road *
Medford vicinity, Jackson County

Constructed - 1922
Timber Through Truss (Queenpost) Covered Bridge
Ownership - Jackson County

ANTELOPE CREEK BRIDGE *

The Antelope Creek Bridge contains a 58-foot housed queenpost truss, modified by the addition of a kingpost. Three of Jackson County's housed spans incorporate the queenpost truss. Now bypassed and unused for vehicular traffic, the Antelope Creek Bridge receives little care and attention. The paint on the ends by the semi-circular portal arches is peeling, and the unpainted sides are weathering. Ribbon openings extend the full length on the side walls under the eaves. Three exposed cantilevered buttresses strengthen the structure.

* In 1987, this bridge was relocated to Eagle Point for pedestrian use, and it was removed from the National Register.

Structure Number 39C409 Constructed - 1922
Battle Creek County Road 4082 Timber Through Truss (Howe) Covered Bridge
Crow vicinity, Lane County Ownership - Lane County

COYOTE CREEK (BATTLE CREEK) BRIDGE
(Also, Swing Log Bridge)

The Coyote Creek Bridge goes by many names, including Battle Creek Bridge (after the road and creek name) and, less often, Swing Log Bridge. The source and meaning of the name Swing Log are unknown but intriguing, nonetheless. This structure contains a 60-foot housed Howe truss. The principal design elements are the housed buttresses, ribbon openings under the eaves, and rectangular portal arches. The Coyote Bridge is on the route of the original Territorial Road (1851). The structure still serves vehicular traffic, although trucks are prohibited.

Structure Number 39C242
Layng County Road 2542
Cottage Grove vicinity, Lane County

Constructed - 1925
Timber Through Truss (Howe) Covered Bridge
Ownership - Lane County

ROW RIVER (CURRIN) BRIDGE

Named after an early pioneer family in the area, the Currin Bridge was constructed in 1925 and is a 105-foot housed Howe truss. An earlier covered bridge at this site was completed in 1883 by veteran bridge builder, Nels Roney. The Currin Bridge has the distinction of being Oregon's only span with white-painted portals and red sides. Its architectural design is the standard Lane County covered bridge. The bridge has been bypassed and is closed to vehicular traffic.

Structure Number 39C446　　Constructed - 1925
Austa County Road　　Timber Through Truss (Howe) Covered Bridge
Richardson vicinity, Lane County　　Ownership - Lane County

WILDCAT CREEK BRIDGE

Constructed in 1925 by Lane County, the bridge crosses Wildcat Creek near its confluence with the Siuslaw River. The housed Howe truss is 75 feet long. The portals have semi-elliptical arches, and ribbon openings extend nearly the length of the side walls under the eaves. A second long narrow opening on the east wall provides a view of oncoming traffic. Decorative brackets are at the ends of the side walls at the eaves. The bridge is open to vehicular traffic, except for trucks. The boat ramp nearby provides access to the Siuslaw River for fishermen.

Structure Number 41C09
Chitwood County Road 427, off the
Corvallis-Newport Highway 33, M.P. 18.11
Chitwood, Lincoln County

Constructed - 1926
Timber Through Truss (Howe)
Covered Bridge
Ownership - Lincoln County

YAQUINA RIVER (CHITWOOD) BRIDGE

Scheduled for replacement and destruction only a few years ago, the Chitwood Covered Bridge is a historic preservation success story in Oregon. Because of strong local sentiment in Lincoln County for saving the bridge, a restoration alternative was chosen to improve the condition of the bridge. CM2M-Hill, an engineering firm from Corvallis, supervised the federally-funded rehabilitation project, completed in 1984. Aubrey Mountain Construction, Eugene, was the contractor. (The above photo shows the restored bridge.) The Chitwood Bridge was originally built in 1926 by Lincoln County, and the design is attributed to A.E. Marvin. The housed Howe truss is 96 feet long. The simple design--semi-elliptical portal arches, flared sides, and ribbon opening under the eaves--is characteristic of the Lincoln County covered bridges. The community of Chitwood, named for Joshua Chitwood, an early resident, was once a thriving railroad town, but is now a ghost town with only a few residents left.

Structure Number 29C211
East Evans Creek County Road
Wimer, Jackson County

Constructed - 1927
Timber Through Truss (Queenpost) Covered Bridge
Ownership - Jackson County

EVANS CREEK (WIMER) BRIDGE

One of only a few Oregon covered bridges located in a city or community center, the Wimer Covered Bridge is an 85-foot housed queenpost truss. The original bridge at this site was erected in 1892. The present bridge was constructed in 1927, but was largely rebuilt in 1962. The bridge was constructed for the county by the Hartman family members of Jacksonville. The bridge includes exposed exterior buttresses, ribbon openings, and semi-circular portal arches. A small park is adjacent to the bridge. It is the only covered bridge in Jackson County open to vehicular traffic. The community of Wimer was established in the 1880s and was named for members of the Wimer family, early residents in the area.

Structure Number 1251
Minnie Ritner Ruiter Wayside off
Kings Valley Highway 19, M.P. 21.03
Kings Valley vicinity, Polk County

Constructed - 1927
Timber Through Truss (Howe)
Covered Bridge
Ownership - Polk County

RITNER CREEK BRIDGE

The Ritner Creek Bridge was the last covered bridge on an Oregon state route. This structure was retired in 1976 and relocated immediately downstream from its original site for use in a wayside park. A one-time levy was passed by the voters of Polk County for the costs of relocation and maintenance. The bridge is closed to vehicular traffic. The 75-foot housed Howe truss was designed and constructed by the State Highway Department with Hamer and Curry, Contractors. The bridge is an example of a state covered bridge standardized design. The portals, once rounded, were cut square to allow larger loads through the bridge.

Structure Number 39C386
Nelson Mountain County Road 3640
Greenleaf vicinity, Lane County

Constructed - 1928
Timber Through Truss (Howe) Covered Bridge
Ownership - Lane County

LAKE CREEK (NELSON MOUNTAIN) BRIDGE
(Also, Nelson Bridge)

Like other bridges in Oregon, this one has several names. Lake Creek refers to the stream that flows under the bridge, while Nelson Mountain refers to the roadway name and nearby geographic feature. Nelson was also the name of a small United States Forest Service camp in the vicinity of the bridge many years ago. Constructed in 1928, the structure contains a 105-foot housed Howe truss. Like many covered bridges, both the upper and lower chords are one-piece old-growth timbers. The window openings are narrow ribbons at the eaves. The portals are truncated rectangular arches. Decorative, solid S-curve brackets carry the eaves at the bridge ends. In 1984, new reinforced concrete abutments, a support bent, and flooring were added to the structure to increase the load capacity of the structure.

Structure Unnumbered
Private Road off the Umpqua Highway 45
Drain vicinity, Douglas County

Constructed - 1929
Timber Through Truss (Howe) Covered Bridge
Ownership - Private

ELK CREEK (ROARING CAMP) BRIDGE

Built in 1929, the Roaring Camp Bridge serves several residents on a private road near Drain. The housed Howe truss is 88 feet long. The bridge is unpainted and clad with 1x6" vertical boards. Ribbon openings run the full length of the upper side walls. The bridge was built by Robert Lancaster. Roaring Camp was a roadhouse, once located in the vicinity of the bridge.

Structure Number 16-5E-24
*Horse Creek County Road **
McKenzie Bridge vicinity, Lane County

Constructed - 1930
Timber Through Truss (Howe) Covered Bridge
Ownership - Lane County

HORSE CREEK BRIDGE *

The Horse Creek Bridge, constructed in 1930, served vehicular traffic until 1968. The bridge is now bypassed and open to pedestrian use only. The predecessor bridge at this location was a 103-foot covered span, built in 1904. The existing bridge contains a 105-foot housed Howe truss. Like other designs in Lane County, the bridge includes both ribbon openings at the eaves on both sides and a window at eye-level on one side for viewing oncoming traffic. The portals are rectangular-arched. Also, the Lane County characteristic "S-curve brackets" are at the corners of the bridge at the eaves.

* This covered bridge was dismantled and placed in storage in 1987.

Structure Number 39C243
Garoutte County Road
Cottage Grove vicinity, Lane County

Constructed - 1930
Timber Through Truss (Howe) Covered Bridge
Ownership - Lane County

MOSBY CREEK (STEWART) BRIDGE

The Stewart Bridge across Mosby Creek is a 60-foot housed Howe truss, constructed in 1930 by Lane County. Because of its deteriorated condition, the covered bridge was bypassed, and a new replacement bridge built. The covered bridge now serves only pedestrian and bicyclist traffic. This short structure has semi-circular portal arches, ribbon openings and decorative S-curve brackets at the eaves on the corners.

Structure Number 12819
Halsey-Sweet Home Highway 212, M.P.
12.94 (Bypassed)
Crawfordsville, Linn County

Constructed - 1932
Timber Through Truss (Howe)
Covered Bridge
Ownership - Linn County

CALAPOOIA RIVER (CRAWFORDSVILLE) BRIDGE

Originally built by the county in 1932, the 105-foot Crawfordsville Bridge became part of a state highway route and remained in state ownership until bypassed in 1963. The bridge is closed to vehicular traffic, but serves as an historical exhibit-in-place in a Linn County park. The community of Crawfordsville was named for Philemon Crawford, who settled in the area in the 1870s. This Linn County structure displays eye-level ribbon openings, instead of the large exposed truss plan of other bridges common to Linn County. The original semi-circular portal arches were modified to allow larger loads through the bridge.

Structure Number 16-9W-25
Deadwood Loop County Road
Swisshome vicinity, Lane County

Constructed - 1932
Timber Through Truss (Howe) Covered Bridge
Ownership - Lane County

DEADWOOD CREEK BRIDGE

Built in 1932 at a cost of $4,814 by Lane County, the Deadwood Creek Bridge is a 105-foot housed Howe truss. The bridge is barricaded and closed to vehicular traffic. Considerable architectural attention has been given to the structure, including false end beams, semi-elliptical portal arches with trim, and large openings along the west elevation.

Structure Number 14721
Jasper-Lowell County Road 6200
Lowell vicinity, Lane County

Constructed - 1936
Timber Through Truss (Howe) Covered Bridge
Ownership - Lane County

FALL CREEK (UNITY) BRIDGE

The Unity Bridge incorporates a 90-foot housed Howe truss. For the most part, this is the standardized Lane County covered bridge design--semi-elliptical portal arches, ribbon openings at the eaves, and ornate brackets at the corners. This bridge does, however, have a full length window on the east side to give motorists a glimpse of oncoming traffic. The eye-level opening is protected by a small projecting roof.

Structure Number 1441
Harris County Road
Wren vicinity, Benton County

Constructed - 1936
Timber Through Truss (Howe) Covered Bridge
Ownership - Benton County

MARYS RIVER (HARRIS) BRIDGE

Completed in 1936, the Harris Bridge is a housed Howe truss, 75 feet long. The Harris Bridge has semi-elliptical portal arches and ribbon openings under the side wall eaves. The siding is board-and-batten style, and the roof is shingled. The Harris Bridge is one of three remaining covered bridges in Benton County and was named for the nearby community of Harris, established in the 1890s.

Structure Number 12948
Camp Morrison Drive County Road 830
Scio vicinity, Linn County

Constructed - 1936
Timber Through Truss (Howe) Covered Bridge
Ownership - Linn County

THOMAS CREEK (HANNAH) BRIDGE

The Hannah Bridge is the youngest of the five covered spans on Thomas Creek in Linn County. The 105-foot housed Howe truss is exposed through the large side openings of the bridge. Very attractive in appearance, the characteristic Linn County covered bridge design includes segmental portal arches, exposed beams at the gable ends, and white board-and-batten cladding. Thomas Creek was named for Frederick Thomas, who obtained a donation land claim on the banks of the stream in 1846. The Thomas Creek (Jordan) Bridge is located two miles east of the Hannah Bridge.

Structure Number 12935
Kelly County Road 622 *
Crabtree vicinity, Linn County

Constructed - 1937
Timber Through Truss (Howe) Covered Bridge
Ownership - Linn County

THOMAS CREEK (WEDDLE) BRIDGE *
(Also, Devaney Bridge)

This bridge goes by two names: Linn County signed the structure "Weddle" Bridge after a nearby farmer, but locals prefer Devaney for another early area resident. This Linn County bridge is closed to vehicular traffic, as a new replacement concrete span was built a quarter-mile away. The old bridge is now used by bicyclists and pedestrians. The housed Howe truss is 120 feet long. The bridge is very similar in design to the other Thomas Creek covered bridges in Linn County.

* Dismantled and placed in storage in 1987.

Structure Number 18-1W-32
Place County Road 6225
Jasper vicinity, Lane County

Constructed - 1938
Timber Through Truss (Howe) Covered Bridge
Ownership - Lane County

FALL CREEK (PENGRA) BRIDGE

Pengra was a station on the Cascade Line of the Southern Pacific Railroad and honors B.J. Pengra, who became General Surveyor of Oregon in 1862. The Pengra Bridge replaced an earlier housed Howe truss. The bridge contains two of the longest timbers ever cut for a bridge in Oregon. Timbers for the single-piece lower chords, measuring 16"x18"x126', were rough-hewn in the woods and finished at the bridge site. The bridge has semi-elliptical arched portals, ribbon openings on the sides, and an eye-level window on one side. The Pengra Bridge is closed to all traffic.

Structure Number 39C118 Constructed - 1938
Goodpasture County Road Timber Through Truss (Howe) Covered Bridge
Vida vicinity, Lane County Ownership - Lane County

McKENZIE RIVER (GOODPASTURE) BRIDGE

One of the most beautiful and photographed covered bridges in Oregon, the Goodpasture Bridge is a popular representative of Oregon's covered bridge heritage. Its location near State Route 126 makes it one of the most visible covered bridges in the state. The bridge was built by Lane County for a cost of $13,000. A.C. Striker was the local county bridge superintendent. Lane County utilized a State Highway Department standardized bridge design. The structure has superb architectural detailing, including ten gothic style louvered windows on each side. Other details include the false end beams and semi-elliptical portal arches. The structure is a 165-foot housed Howe truss. The bridge is named for a pioneer family who settled nearby.

Structure Number 39C174
Wendling County Road
Marcola vicinity, Lane County

Constructed - 1938
Timber Through Truss (Howe) Covered Bridge
Ownership - Lane County

MILL CREEK (WENDLING) BRIDGE

The Wendling Bridge is one of four covered bridges built in 1938 by Lane County. The others are the Pengra, Goodpasture and Ernest bridges. A 60-foot housed Howe truss, the Wendling Bridge was named for George X. Wendling, director of the Booth-Kelly Lumber Company, who established a post office in the 1890s in the small town nearby. Lane County spent only $2,241 to build the bridge, and A.C. Striker was the Lane County bridge superintendent at the time. Architectural treatments include the semi-elliptical portal arches, portal trim boards, and ribbon openings at the eaves.

Structure Number 39C176
Paschelke County Road 1980
Marcola vicinity, Lane County

Constructed - 1938
Timber Through Truss (Howe) Covered Bridge
Ownership - Lane County

MOHAWK RIVER (EARNEST) BRIDGE

The original covered span at this location was erected in 1903 by A.C. Striker and was called the Adams Bridge. Named for another longtime local resident, the 1938 replacement bridge contains a 75-foot housed Howe truss. The bridge is freshly painted and appears to be in good condition. Another standard Lane County design, a distinctive feature of the Earnest Bridge is the small, hooded opening on one elevation of the bridge for motorist visibility.

Structure Number 12037
North Fork Yachats River County Road
Yachats vicinity, Lincoln County

Constructed - 1938
Timber Through Truss (Queenpost) Covered Bridge
Ownership - Lincoln County

NORTH FORK YACHATS RIVER BRIDGE

This bridge is one of the shortest covered bridges in the state at only 42 feet long. Built in 1938, it was the last covered span constructed by veteran bridge builder Otis Hamer. This span is one of four remaining covered bridges in Lincoln County, and only this bridge and the restored Chitwood Bridge are open to vehicular traffic. The flared sides of the Lincoln County covered bridges result from the enclosed buttresses. The arched portals and short ribbon openings under the eaves provide interior illumination.

Structure Number 12876
Fish Hatchery Drive County Road 648
Crabtree vicinity, Linn County

Constructed - 1939
Timber Through Truss (Howe) Covered Bridge
Ownership - Linn County

CRABTREE CREEK (LARWOOD) BRIDGE

Located at the confluence of Roaring River and Crabtree Creek, the Larwood Bridge carries the historical name of the community. William T. Larwood opened the post office at a site in the vicinity of the bridge in 1893. The Larwood Bridge is one of three covered bridges across the Crabtree Creek in Linn County. The 105-foot housed Howe truss exhibits the common Linn County design of exposed truss side openings. The bridge is open to vehicular traffic and is adjacent to the Larwood Wayside Park.

Structure Unnumbered
Private Road
Westfir vicinity, Lane County
Constructed - 1944
Timber Through Truss (Howe) Covered Bridge
Ownership - Private

NORTH FORK MIDDLE FORK WILLAMETTE RIVER (OFFICE) BRIDGE

The most massive and longest of Oregon's covered bridges, this 180-foot housed Howe truss was constructed by the Westfir Lumber Company using triple timber beams to afford the strength necessary to carry heavy logging trucks. The bridge connects the lumber mill with the office (hence the common name of the bridge). The structure is one of only two covered bridges in Oregon built with triple truss members. A distinction of the bridge is the covered walkway on the side of the bridge, separate from the roadway. Westfir was established in 1923 as a company town by the Western Lumber Company. The privately-owned bridge is closed to public traffic.

Structure Number 19-1W-23
Jasper-Lowell County Road (Bypassed)
Lowell vicinity, Lane County

Constructed - 1945
Timber Through Truss (Howe) Covered Bridge
Ownership - Lane County

MIDDLE FORK WILLAMETTE RIVER (LOWELL) BRIDGE

Bypassed in 1980, the Lowell Bridge is the only covered bridge in Oregon spanning a portion of a reservoir. Built in 1945, the Lowell Bridge was raised six feet and the roadway rebuilt in 1953 in anticipation of the pool formed between Dexter Dam and Lookout Point Dam. This is the second covered bridge at Lowell. Its predecessor was a Nels Roney-built bridge constructed in 1907. The current Lowell Covered Bridge is a housed Howe truss, 165 feet in length. The generous width of the bridge allows a roadway of 24 feet and is typical of covered bridges built in the 1940s and after. Distinctive features are the false end beams, semi-elliptical portal arches and two arched openings on the east elevation. The bridge is closed to vehicular traffic.

Structure Number 14025
High Deck County Road 913
Cascadia, Linn County

Constructed - 1945
Timber Through Truss (Howe) Covered Bridge
Ownership - Linn County

SOUTH FORK SANTIAM RIVER (SHORT) BRIDGE
(Also, Whiskey Butte Bridge)

Originally named for the nearby topographic feature, Whiskey Butte, the bridge was later named for long-term area resident, Gordon Short. The housed Howe truss is 105 feet in length and has the distinctive open feeling of most of Linn County's covered bridges. The open truss provides increased height visibility to the traveler and keeps wind resistance to a minimum. Familiar design features also include segmental portal arches, exposed gable end beams, and white board-and-batten siding.

Structure Number 21-2W-24A
Government County Road 2440
Dorena vicinity, Lane County

Constructed - 1949
Timber Through Truss (Howe) Covered Bridge
Ownership - Lane County

ROW RIVER (DORENA) BRIDGE
(Also, Star Bridge)

The Dorena Bridge spans the Row River beyond the upper end of the Dorena Reservoir. The 105-foot housed Howe truss structure was built in 1949, in conjunction with the completion of Dorena Dam. The original townsite of Dorena, established at the turn of the century, was inundated by the waters of the reservoir. The bridge is also referred to as the Star Bridge because of its proximity to the Star Ranch, once a large Lane County private estate. The bridge was bypassed in 1974 and is closed to vehicular traffic. Design features include false end beams, rectangular portal arches, and ribbon openings at the roof line.

Structure Number 45005-17
Irish Bend County Road (Bypassed)
Monroe vicinity, Benton County

Constructed - 1954
Timber Through Truss (Howe) Covered Bridge
Ownership - Benton County

WILLAMETTE SLOUGH (IRISH BEND) BRIDGE

The 1954-constructed Irish Bend Bridge is barricaded and bypassed. A replacement structure is adjacent to the covered bridge and provides service to the farms in the area. The covered bridge is a 60-foot housed Howe truss. The portals are truncated rectangular arches, and there are ribbon openings under the side wall eaves. The span has long, trestled approaches. The bridge was built by Benton County and is a plan from the 1920s. Benton County has tried unsuccessfully to find a new owner to relocate and repair the bridge.

Structure Unnumbered
Milo Academy Private Road
Days Creek vicinity, Douglas County

Constructed - 1962
Steel Through Girder Covered Bridge
Ownership - Private

SOUTH UMPQUA RIVER (MILO ACADEMY) BRIDGE

The Milo Academy Bridge is the replacement for a covered timber structure erected at this location in 1920. The present 100-foot steel through plate girder span, constructed in 1962, was housed in response to requests from local residents to recreate the effect of the original covered span. The Milo Academy Bridge is one of only two covered bridges in the Oregon Covered Bridges Thematic Group which do not have a timber truss support. The bridge has rectangular portals and formally-placed rectangular side openings. The roof is metal, and the bridge is clad with vertical wood siding. Milo was established as a post office in 1923 and was named after Milo, Maine. The private bridge provides access to a Seventh Day Adventist school.

Structure Unnumbered
Bowery Lane Private Road
Bend vicinity, Deschutes County

Constructed - 1963
Timber Deck Girder Covered Bridge
Ownership - Private

SWALLEY CANAL (ROCK O' THE RANGE) BRIDGE

The Rock O' the Range Bridge is the only covered bridge in eastern Oregon and is a recently constructed one. Although constructed of wood, it is not supported by a truss. Instead, it is a housing on a timber deck girder structure. The bridge was built for a private subdivision by Maurice Olson, a local contractor. It is an architecturally interesting structure, even though its design lacks a historic tradition in Oregon. This bridge is 42 feet long.

Structure Number 39C123
King Road West County Road
Rainbow, Lane County

Constructed - 1966
Timber Through Truss (Howe) Covered Bridge
Ownership - Lane County

McKENZIE RIVER (BELKNAP) BRIDGE

Although the Belknap Bridge is the youngest bridge (1966) in the Oregon Covered Bridges Thematic Group, it succeeds three earlier covered spans at this location. Previous bridges were constructed in 1890, 1911, and 1938. The existing bridge was designed by the Oregon Bridge Engineering Corporation (OBEC), Eugene, for Lane County. Louvered arch windows were added in 1975 to the south side to give interior illumination. The portals are rectangular-arched. The name Belknap refers to early pioneers along the McKenzie River. R.S. Belknap developed Belknap Springs, while his son J.H. Belknap had an interest in the toll road that was built over McKenzie Pass in the early 1870s. This structure has a span of 120 feet.

OTHER HISTORIC BRIDGES

Nine highway bridges in Oregon have been individually recognized as historically significant. Six of the bridges have been formally determined eligible for the National Register, while the other three are listed on the National Register. With the exception of the two Columbia River crossings, Interstate Northbound Bridge (1917) and Lewis and Clark Bridge (1930), each of these bridges received their National Register status by separate actions.

The ODOT, acting through the Federal Highway Administration, requested National Register eligibility status on the following four bridges proposed for modification or replacement: Willamette River (Steel), Oswego Creek, Alsea Bay (Waldport), and Thomas Creek (Jordan). Two federal agencies requested historic opinions on highway bridges for planning purposes: United States Soil Conservation Service, Rock Creek (Olex) Bridge; and United States Forest Service, North Umpqua River (Mott) Bridge.

The S.W. Vista Avenue Viaduct, a 1926 reinforced concrete arch structure in Portland, was listed on the National Register in April 1984. That nomination was sponsored by a volunteer group of concerned residents, the Vista Bridge Light Brigade.

The two Columbia River bridges, Interstate Northbound and Lewis and Clark, were inventoried and evaluated for historic significance in conjunction with the Washington historic bridges survey. The Washington survey was conducted by the State Office of Archeology and Historic Preservation, in cooperation with the Washington State Department of Transportation and the Historic American Engineering Record. That survey concluded in a thematic National Register nomination, "Historic Bridges/Tunnels in Washington State," which included the two structures adjoining Oregon and Washington. The bridges were listed on the National Register in July 1982.

The nine bridges are summarized below and shown on the following photo-description pages.

THE OTHER HISTORIC BRIDGES

Bridge Name	Bridge Number	County	Type	Date
Rock Creek (Olex)	—	Gilliam	Steel Pony Truss* (Half-Hip Pratt)	Ca. 1905
Willamette River (Steel)	2733	Multnomah	Steel Through Truss (Pratt) Vertical Lift	1912
Columbia River (Interstate Northbound)	1377A	Multnomah	Steel Through Truss (Pennsylvania-Petit) Vertical Lift	1917
Oswego Creek	409	Clackamas	Reinforced Concrete Deck Arch	1920
S.W. Vista Avenue Viaduct	25B36	Multnomah	Reinforced Concrete Deck Arch	1926
Columbia River (Lewis and Clark)	433/1	Columbia	Steel Through Truss (Cantilever)	1930
Alsea Bay (Waldport)	1746	Lincoln	Reinforced Concrete Through Tied Arch	1936
North Umpqua River (Mott)	4712-000-0.1	Douglas	Timber Deck Arch	1936
Thomas Creek (Jordan)	12958	Linn	Timber Through Truss (Howe) Covered Bridge	1937

*Pin-connected truss.

Structure Unnumbered
Private Road
Olex vicinity, Gilliam County

Constructed - Ca. 1905
Steel Pony Truss (Half-Hip Pratt)
Ownership - Private

ROCK CREEK (OLEX) BRIDGE

This pony truss was built by the Pacific Iron Works of Portland to replace an earlier timber bridge and is one of only three known examples of the half-hip Pratt truss in Oregon. The truss members are pin connected. In proximity to the bridge are the Olex School (1903), Crum Gristmill (1883), and an old wagon road. The Rock Creek Bridge was built about 1905, based on available records. The bridge was determined eligible for the National Register in November 1975.

Structure Number 2733
Pacific Highway West 1W, M.P. 0.36
Portland, Multnomah County

Constructed - 1912
Steel Through Truss (Pratt) Vertical Lift
Ownership - Union Pacific
and Southern Pacific railroads

WILLAMETTE RIVER (STEEL) BRIDGE

Built jointly by the Oregon Railway and Navigation Company and the Union Pacific Railroad, the Steel Bridge was the largest telescoping bridge in the world at the time of its opening. The telescoping function of the central span, a 211-foot steel, through Pratt truss, double vertical lift span, makes the Steel Bridge an important example of a rare engineering design. The lower railway deck can be raised for the passage of small vessels without disturbing automobile traffic on the upper deck. For larger vessels, both spans can be raised. It is believed to be the world's only double-lift span that can raise its lower deck independently of the upper deck. Opening both decks allows for 163 feet total clearance. The bridge was designed by Waddell and Harrington, consulting engineers from Kansas City, and the railroad engineers. The bridge was built over a two-year period at a cost of $1.7 million. There are two secondary steel through Pratt truss spans, each 290 feet long. The structure has little decorative embellishment, other than a wrought iron woven lattice railing. This structure is located near the site of the first Steel Bridge (1888). The 1912 structure was determined eligible for the National Register in April 1980.

Structure Number 1377A
Pacific Highway 1 (I-5), M.P. 308.38
Portland vicinity, Multnomah County
(Oregon) and Vancouver, Clark County (Washington)

Constructed - 1917
Steel Through Truss (Pennsylvania-Petit)
Vertical Lift
Ownership - State of Oregon

COLUMBIA RIVER (INTERSTATE NORTHBOUND) BRIDGE

This structure, listed on the National Register in July 1982, was a major engineering and financial accomplishment, being the first highway bridge across the Columbia River to connect Oregon and Washington. The bridge was designed by Harrington, Howard, and Ash, a Kansas City engineering firm, and was constructed for a cost of $1,683,000. The main span is a 279-foot steel through truss vertical lift span of the Pennsylvania-Petit type. There are ten steel through Pennsylvania truss secondary spans, ranging in length from 266 to 531 feet, providing a total structure length of 3,538 feet. The matching parallel bridge structure immediately to the west (downstream) of the 1917 bridge was constructed in 1958.

Structure Number 409
Oswego Highway 3, M.P. 6.76
Lake Oswego, Clackamas County

Constructed - 1920
Reinforced Concrete Deck Arch
Ownership - State of Oregon

OSWEGO CREEK BRIDGE
(Also, Sucker Creek Bridge)

The Oswego Creek Bridge, designed by Conde B. McCullough and built by the Pacific Bridge Company of Portland, was originally known as the Sucker Creek Bridge. Public opinion forced a name change in the creek, and hence the present name of the bridge. The bridge is a local landmark and has considerable artistic treatment. Because of the structure's prominence on the old Pacific Highway, the design was deliberately made rather pretentious. Two types of cement were used to provide a striking contrast in the surface. The structure is 300 feet long. The main span, 130 feet in length, is a three-radius shaped, open-spandrel rib arch of reinforced concrete. The spandrel columns have semi-circular arch fascia curtain walls, and the railings are a series of small curved arch openings. A major addition to the bridge was completed on the downstream side of the bridge in 1983 to provide additional travel lanes and safer conditions. The structure addition is a prestressed slab span and is sensitive in design to the old arch bridge. (The above view was taken prior to the project addition.) The bridge was determined eligible for the National Register in December 1979.

Structure Number 25B36
S.W. Vista Avenue/S.W. Canyon Boulevard
Portland, Multnomah County

Constructed - 1926
Reinforced Concrete Deck Arch
Ownership - City of Portland

S.W. VISTA AVENUE VIADUCT

The main span of this monumental structure is a 248-foot open-spandrel, rib-type reinforced concrete deck arch. The S.W. Vista Avenue Viaduct is both a highly attractive bridge and a popular local landmark. Great attention was given to decorative detail by its designer, Portland city bridge engineer Fred T. Fowler, evidenced by the ornamental lampposts, pylons, pebble-dashed inset panels, scrollwork, and the spindle-type balustrade railing. Four pedestrian balconies offer vistas of Mt. Hood and downtown Portland. The bridge is a designated Portland Historical Landmark and was listed on the National Register in April 1984. In 1988, new replicas of the ornamental lampposts (that had been removed over time) and other details were restored to this bridge.

Structure Number 433/1
Washington State Route 433 at the
Lower Columbia River Highway 2W, M.P. 48.92
Rainier, Columbia County (Oregon) and
Longview, Cowlitz County (Washington)

Constructed - 1930
Steel Through Truss (Cantilever)
Ownership - State of Washington

COLUMBIA RIVER (LEWIS AND CLARK) BRIDGE
(Also, Longview Bridge)

The Lewis and Clark Bridge is an impressive steel through cantilever truss structure. The entire structure is 1.6 miles in length (8,288 feet) including approaches. The 2,722-foot cantilever truss has a main span which measures 1,200 feet and rises 195 feet above the Columbia River channel. When constructed, it was the longest cantilever span in the United States. The chief designer was Joseph B. Strauss of Chicago, who later designed the Golden Gate Bridge (1937) in San Francisco. The general contractor was the Bethlehem Steel Company. The bridge was privately built by the Columbia River-Longview Company in 1930 and purchased in 1947 by the Washington State Department of Transportation. In July 1980, the bridge was officially renamed the Lewis and Clark Bridge. The bridge was listed on the National Register in July 1982.

Structure Number 1746
Oregon Coast Highway 9, M.P. 155.54
Waldport, Lincoln County

Constructed - 1936
Reinforced Concrete Through Tied Arch
Ownership - State of Oregon

ALSEA BAY (WALDPORT) BRIDGE

The main spans of this 3,011-foot structure are three reinforced concrete through tied arches, 154, 210, and 154 feet in length. Three 150-foot concrete deck arch secondary spans are on either side of the main spans. Thirty-two concrete deck girder approach spans also complement the structure. Decorative features include ornamental railing and spandrel post brackets, fluted entrance pylons, obelisk spires at the arch portals, and a pedestrian observation plaza. Significant for its impressive design and ornate treatment, the bridge was one of five major structures built during Oregon's WPA-financed Coast Bridges project, all designed by Conde B. McCullough, State Bridge Engineer, and completed in 1936. The overall appearance of the Alsea Bay Bridge is one of grace, rhythm, and harmony with the marine setting. The Alsea Bay Bridge is the largest of the reinforced concrete tied-arch bridges by McCullough and is considered by some experts to rank among the finest examples of concrete bridge construction in America. The structure was determined eligible for the National Register in March 1981. The Alsea Bay Bridge is scheduled for displacement, and final planning is underway for the replacement structure.

Structure Number 4712-000-0.1
Steamboat Ranger Station Road
Steamboat vicinity, Douglas County

Constructed - 1936
Timber Deck Arch
Ownership - Umpqua National Forest,
United States Forest Service

NORTH UMPQUA RIVER (MOTT) BRIDGE

One of only two known timber deck arches in Oregon, this 135-foot braced-spandrel, three-hinged arch span was constructed by the Civilian Conservation Corps and is a rare bridge type. The bridge was dedicated as the Mott Bridge in honor of nationally-known author and sportsman, Major Mott, who established a fishing camp near the bridge site. The bridge currently provides access to the Steamboat Ranger Station. The bridge was determined eligible for the National Register in November 1980.

Structure Number 12958　　Constructed - 1937
Jordan County Road 829 *　　Timber Through Truss (Howe) Covered Bridge
Lyons vicinity, Linn County　　Ownership - Linn County

THOMAS CREEK (JORDAN) BRIDGE *

A prominent local landmark and an attractive example of covered bridges, the Jordan Bridge is a 90-foot housed Howe truss. The bridge takes its name from the nearby community of Jordan, named for the Jordan Valley in Israel. The Jordan Bridge is one of Linn County's seven distinctive covered bridges with large side openings. The portal arches were originally rounded or segmental, but have been altered to a truncated rectangular shape, expanding clearance for large loads. The bridge was determined eligible for the National Register in February 1984.

* Relocated to Pioneer Park in Stayton for pedestrian use, this bridge was rededicated in June 1988.

CONCLUSION

Rogue River Bridge (1931), Oregon Coast Highway, Gold Beach, Curry County

CONCLUSION

The historic highway bridges study was a comprehensive inventory and evaluation of the highway bridges in Oregon. The 145 significant structures distributed throughout the state represent the best in Oregon's bridge-building tradition, combining the 68 study-identified bridges and the 77 highway bridges previously determined eligible for or listed on the National Register.

Identifying the National Register-eligible highway bridges is a major step toward resolving the conflicting interests of modern transportation needs and historic preservation. Because of their historic importance, these bridges will require special planning considerations in accordance with federal and state historic preservation laws and regulations, if and when replacement or major rehabilitation is proposed. Their National Register eligibility identifies them as worthy of preservation, even if it may not always be possible or desirable to save every bridge. The historic bridges study provides information on the historically-significant structures and allows better planning and development of projects. Of the approximate 7,000 highway bridges in the state, only about two percent are of historic significance. The remainder, the vast majority of Oregon's highway bridges, are non-historic and can undergo normal replacement planning as required.

The historic highway bridges study provides useful data for the development of an ODOT bridge preservation priority plan. The preservation priority plan will address the complex issues relating to the costs and feasibility of preserving specific historic bridges.

Every effort was made in the study to identify all highway bridges in the state with historic significance, but additional bridges may be identified in the future and added to the list of National Register-eligible bridges. Several bridges were discovered during field inspections that were not listed on the ODOT's computerized list. These bridges were incorporated into the study. Other bridges may, likewise, be discovered in the future and will need to be evaluated for their historic significance.

For the most part, the study-identified bridges were evaluated as individually significant at the statewide level and not as components in other thematic groups of bridges. The study bridges were not considered as components of nontransportation historic districts, ensembles, or groupings. National Register nominations or requests for a determination of eligibility may be developed that include bridges which are not distinguished in themselves, but contribute thematically or historically to other properties. Consequently, additional highway bridges may be defined as National Register eligible in the future, although the number is not expected to be large.

The Department of Transportation will periodically update the study. This will include a reevaluation of the historic importance of the reserve bridges, as historic and other early non-historic bridges are lost, reducing the number of certain types of bridges. The update will also provide an opportunity to identify bridges which are locally significant or included in thematic groups or historic districts of nonengineering or transportation resources.

FIGURE 24. The Coquille River Bridge (1922) at Coquille, shown in this early photograph, is a remaining example of a large swing span in Oregon. Once a common moveable type across Oregon's navigable waterways, the swing span is now an obsolete form. Although being replaced by a new bridge downstream, the old Coquille River Bridge will be retained and maintained for pedestrian use by the Oregon Department of Transportation.

There are many things that make up the fabric of our history, and some things are more fundamental than others. One of the oldest engineering works devised by man, bridges are basic to our civilization. The majority of our bridges in Oregon are merely "work horses"--essential, utilitarian and not usually appreciated (unless they are not there). But, like all elements in the built environment, bridges also bear the imprint of creativity and exhibit the same range and development as other products of our culture.

Bridge architecture and engineering occasionally go beyond the strictly utilitarian into the magnificent and sublime, and become works of art. It is hoped that many of the bridges in this report fit into that category. The history of bridge building is rich in tradition and innovation, as new needs, materials, and technologies helped develop longer, lighter, more efficient structures--the same goal toward which today's engineers also strive.

There has always been an interest in bridges and their preservation, especially for some types of bridges. Until lately that interest usually centered only on the small and the picturesque (covered bridges) and the very large and aesthetic (suspension bridges). Public interest is now beginning to encompass the middle range of bridges, too.

Because of increased federal and state appropriations for bridge replacements and a nationwide concern for improvement of bridges, the rapid destruction of bridges is a critical historic preservation issue. Unfortunately, bridges pose difficult problems in preservation because of their importance in modern transportation systems, safety, liability, the lack of feasible adaptive reuses, and the costs of maintenance. It is through studies like this in Oregon and across the nation that the magnitude of the historic resource will be made known. The future for bridge preservation now depends on new legislation and sources of funding to encourage the rehabilitation and continued use of historic bridges and the discovery of new and creative adaptive reuses for bridges.

It is hoped that many of Oregon's historic bridges will be preserved for future generations to appreciate and study. It would be a sad situation indeed if, through the lack of foresight, the historic bridges were lost and existed only in documents and memories.

FIGURE 25. *The Old Rhinehart Bridge (1922) across the Grande Ronde River in Union County is bypassed and abandoned-in-place, one of many methods for preserving historic bridges in Oregon.*

BIBLIOGRAPHY

Yaquina Bay Bridge (1936), Oregon Coast Highway, Newport, Lincoln County

BIBLIOGRAPHY

The references below include only major and general sources of information. The list is not exhaustive of all consulted sources, and for the most part, excludes references to specific bridges.

Ammann, O.H. "Advances in Bridge Construction." *Civil Engineering,* August 3, 1933: 428-32.

Billington, David P. "History and Esthetics in Concrete Arch Bridges." *Journal of the Structural Division,* American Society of Civil Engineers 103, November 1977: 2129-43.

Black, Archibald. *The Story of Bridges.* New York: McGraw-Hill, 1936.

"Bridges--A Special Issue." *Society for Industrial Archeology Newsletter,* Volume 8, Numbers 1 and 2, January and March 1979.

Chamberlin, William P. "Historic Bridges: Criteria for Decision Making." Washington, D.C.: Transportation Research Board, 1983.

Cockrell, Nick, and Bill Cockrell. *Roofs Over Rivers: A Guide to Oregon's Covered Bridges.* Beaverton, Oregon: The Touchstone Press, 1978.

Comp, T. Allan, and Donald Jackson. "Bridge Truss Types: A Guide to Dating and Identifying." American Association for State and Local History. Technical Leaflet 95. *History News,* Volume 32, No. 5, May 1977.

Condit, Carl W. *American Building Art: The Nineteenth Century.* New York: Oxford University Press, 1960.

_____. *American Building Art: The Twentieth Century.* New York: Oxford University Press, 1961.

Corning, Howard M. (ed.). *Dictionary of Oregon History.* Portland: Binfords and Mort, 1956.

_____. *Willamette Landings: Ghost Towns of the River.* Portland: Oregon Historical Society, Second Edition, 1973.

Deibler, Dan G. *A Survey and Photographic Inventory of Metal Truss Bridges in Virginia, 1856-1932,* Volumes I-IV. Charlottesville, Virginia: Virginia Highway and Transportation Research Council, 1975.

Deloney, Eric (ed.). "Historic Bridge Bulletin," *Society for Industrial Archeology Newsletter,* Summer 1984, Number 1.

Dicken, Samuel N., and Emily F. Dicken. *The Making of Oregon: A Study in Historical Geography.* Portland: Oregon Historical Society, 1979.

Elling, Rudolf E., and Gayland B. Witherspoon. *Metal Truss Highway Bridge Inventory.* Columbia: South Carolina Department of Highways and Public Transportation, 1981.

Fore, George. *North Carolina Metal Bridges: An Inventory and Evaluation.* Raleigh, North Carolina: North Carolina Division of Archives and History and North Carolina Division of Highways, 1979.

Fraser Design. *Wyoming Truss Bridge Survey.* Cheyenne: Wyoming State Highway Department, 1984.

Galli, G. "100 Years of Construction News--Events that Shaped the Future." *Engineering News Record* 192, April 1974: 455.

Geis, Joseph. *Bridges and Men.* New York: Grosset and Dunlap, 1966.

Georgia Department of Transportation and Georgia Department of Natural Resources. *Historic Bridge Survey.* Atlanta, 1981.

Guedes, Pedro (ed.). *Encyclopedia of Architectural Technology.* New York: McGraw-Hill Book Company, 1979.

Hayden, Martin. *The Book of Bridges.* New York City: Galahad Books, 1976.

Historic American Highways. Washington D.C.: American Association of State Highway Officials, 1953.

Hoyt, Jr., Hugh M. *The Good Roads Movement in Oregon: 1900-1920.* Unpublished thesis, Department of History, Graduate School of the University of Oregon, 1966.

Jackson, Donald C. "Railroads, Truss Bridges and the Rise of the Civil Engineer." *Civil Engineering* 47, October 1977: 97-101.

_____, and Shanahan, Nancy C. *et al.* "Saving Historic Bridges." Information from the National Trust for Historic Preservation, Sheet 36. Washington, D.C.: 1984.

Kemp, Emory L. *West Virginia's Historic Bridges.* Charleston: West Virginia Department of Culture and History and West Virginia Department of Highways, 1984.

Kentucky Department of Transportation. *A Survey of Truss, Suspension, and Arch Bridges in Kentucky.* Frankfort: Kentucky Bureau of Highways, 1982.

Ketchum, Milo S. *The Design of Highway Bridges and the Calculation of Stresses in Bridge Trusses.* New York: Engineering News Publishing Co., 1909.

_____. *The Design of Highway Bridges.* New York: McGraw-Hill Book Company, Inc., 1920.

Lindenthal, Gustav. "Some Aspects of Bridge Architecture," *Scientific American,* November 1921.

Loy, William G. *Atlas of Oregon.* Eugene: University of Oregon Books, 1976.

MacColl, E. Kimbark. *The Shaping of a City: Business and Politics in Portland, Oregon, 1885-1915.* Portland: Georgian Press Company, 1976.

_____. *The Growth of a City: Power and Politics in Portland, Oregon, 1915 to 1950.* Portland: Georgian Press Company, 1979.

McArthur, Lewis A. *Oregon Geographic Names.* Fifth Edition. (Edited by Lewis L. McArthur.) Portland: Oregon Historical Society, 1982.

McCullough, Conde B. *Economics of Highway Bridge Types.* Chicago: Gillette Publishing Company, 1929.

_____, and Edward S. Thayer. *Elastic Arch Bridges.* New York: J. Wiley and Sons, 1931.

_____, and John R. McCullough. *The Engineer at Law: A Resume of Modern Engineers Jurisprudence.* Ames, Iowa: The Collegiate Press, 1946.

Mock, Elizabeth B. *The Architecture of Bridges.* New York: The Museum of Modern Art, 1949.

Moore, Ralph T. *Historic Transportation System in Oregon: Present and Future Needs.* Based on an Engineering Analysis for the Legislation Interim Committee for the Study of Highways, Roads and Street Needs, Revenue and Taxation. Salem, 1948.

Nelson, Lee H. *Oregon Covered Bridges.* Portland: Oregon Historical Society, 1976.

Newlon, Howard. *Criteria for Preservation and Adaptive Use of Historic Highway Structures.* Charlottesville, Virginia: Virginia Highway and Transportation Research Council, 1978.

Ohio Department of Transportation. *The Ohio Historic Bridge Inventory Evaluation and Preservation Plan.* Columbus: 1983.

Oregon Historical Society. *Oregon Historical Quarterly.* Portland: Various Issues.

Oregon State Highway Commission. *Biennial Report.* Salem: Oregon State Highway Department, reports covering the years 1913 to 1941.

Oregon State Highway Division. *Bridge Log.* Salem: Oregon Department of Transportation, 1983.

Pierce, Louis F. "Esthetics in Oregon Bridges--McCullough To Date." Portland: American Society of Civil Engineers, 1980.

Plowden, David. *Bridges: The Spans of North America.* New York: Viking Press, 1974.

Potter, Elizabeth Walton. "Oregon Covered Bridges," Thematic group nomination to the National Register of Historic Places. Salem: State Historic Preservation Office, 1979.

Quivik, Fredric L. *Historic Bridges in Montana.* Helena: State of Montana Department of Highways, 1982.

Smith, Dwight A. *Columbia River Highway District: Nomination of the Old Columbia River Highway in the Columbia Gorge to the National Register of Historic Places.* Salem: Oregon Department of Transportation, 1984.

Soderborg, Lisa. *Historic Bridges in Washington State.* Olympia: Washington State Office of Archeology and Historic Preservation, 1981.

Staehli, Alfred M. *Willamette River Bridge 5.1 Draw Span.* Prepared for the Historic American Engineering Record. (Contains history of the Spokane, Portland, and Seattle Railway by Lewis L. McArthur.) Portland: 1985.

Tyrrell, Henry G. *History of Bridge Engineering.* Chicago: By the Author, 1911.

United States Department of Transportation, Federal Highway Administration. *America on the Move: The Story of the Federal-Aid Highway Program.* Washington, D.C.: Government Printing Office, 1962.

_____. *A Nation in Motion: Historic American Transportation Sites.* Washington, D.C.: Government Printing Office, 1976.

Waddell, J.A.L. *Bridge Engineering.* New York: John Wiley and Sons, 1916.

Watson, Ralph. *Casual and Factual Glimpses at the Beginning and Development of Oregon's Roads and Highways.* Salem: Oregon State Highway Department, about 1954.

Watson, Wilbur J. *Bridge Architecture.* New York: William Helburn, Inc., 1927.

Whitney, Charles S. *Bridges: Their Art, Science and Evolution.* New York: Crown Publishers, Inc., 1983 Reprint of 1929 book.

GLOSSARY

Burnside Bridge (1926), Willamette River, Portland, Multnomah County

GLOSSARY

Abutment	A substructure supporting the end of a single span or the extreme end of a multi-span structure and, in general, supporting the approach embankment.
Approach	The passageway structure from the roadbed onto the bridge structure.
Aqueduct	Bridge which carries a canal or a water conduit.
Balustrade	A railing or parapet consisting of a handrail on balusters (vertical support members).
Bascule	A moveable bridge in which the roadway deck is counterbalanced by a weight and swings upward like the hinged cover of a box. A bascule bridge may be single-leaf (one hinge) or double-leaf (two hinges).
Beam	A bridge type in which the roadway deck is supported by transverse members framed into the outer longitudinal structural members.
BPR	United States Bureau of Public Roads, the predecessor organization to the Federal Highway Administration, United States Department of Transportation.
Bracket	An overhanging member that projects from the structure, designed to support a vertical load, usually a sidewalk, or sometimes only the railing.
Bridge	A structure which provides continuous passage over a body of water, roadway, or valley. Generally, a bridge carries a pathway, road, or railroad, but it may also carry power transmission lines. (Bridges in this study were restricted to those of 20 feet or longer and which carry public highways, roads, or streets.)
Bush-hammered	A treatment to concrete surfaces with a steel-plated instrument which results in a textured surface instead of a flat, smooth surface. This technique is also referred to as pebble-dash.
Buttress	An abutting pier which strengthens a wall, sometimes taking the thrust of an inner pier.
Ca.	Circa or estimated date.
Cantilever	Any rigid structural member projecting from a vertical support, especially one in which the projections is great with relation to the depth, so that the upper part is in tension and the lower part in compression.
Chord	A main outer structural member of a truss.
Compression members	Generally stiff, heavy posts composed of channel and I-bars which withstand pressure that tends to push them together.
Continuous structure	A generally long bridge in which the structure is supported by more than two piers, but still distributes stress over the entire structure.

Counter	The adjustable diagonal in a truss, not liable to stress except upon partial application of live loads.
Covered bridge	A structure, usually timber, with a housing to protect the truss and other members from the elements.
Culvert	A drain or channel crossing under a road. Generally, culverts are small and constructed of steel or concrete. When they become large and are not continuous under the waterway, they become slab span bridges.
Curtain wall	The outer members in a bridge or approach span, particularly when the configuration differs from the inner members; also called fascia.
Deck	The roadway surface. Also, a bridge type with the roadway atop the bridge framework.
Dentils	A series of small projecting rectangular blocks, especially under a cornice or other overhanging structure, used for ornamentation.
Fascia	See Curtain wall.
FHWA	Federal Highway Administration, an agency in the U.S. Department of Transportation.
Girder	A bridge type in which the roadway deck is supported by one or more longitudinal structural members.
Gussett plate	A triangular piece which stiffens an angular meeting of two or more members in a framework, frequently found on metal truss bridges and secured with rivets.
Lattice	A vintage railing type that consists of a system of crosshatched diagonals with no verticals.
Moveable bridge	A bridge type which opens to allow additional vertical clearance for water navigation.
National Register	The National Register of Historic Places, maintained by the National Park Service, United States Department of Interior, is the list of the nation's cultural resources worthy of preservation. This list contains districts, historic and archeological sites, buildings, structures, and objects of national, state, and local significance.
ODOT	Oregon Department of Transportation.

OSHD	Formerly the Oregon State Highway Department, this state agency became the Oregon State Highway Division within the Oregon Department of Transportation in 1969.
Overcrossing or overpass	A bridge structure where the principal or subject transportation facility is the upper roadway (of two roadway levels).
Parapet	A low retaining wall or railing.
Pier	A structure which supports the ends of the spans of a multi-span superstructure at an intermediate location between its abutments.
Pin connected	A feature of early truss construction in which the truss members were joined by steel pins or bolts.
Polygonal	Having many angles.
Pony truss	A low through truss that has no overhead or enclosing truss work. (The word "pony" indicates a scale of measurement, something smaller than standard.)
Portal	The entrance to a bridge, especially a through truss or arch.
Portal message	A plaque mounted above the entrance portal of a "through" bridge, indicating the company that constructed the bridge, the date of construction, etc.
Rivet connected	A rigid connection of steel bridge members, which replaced pin-connections. The rivet connection increased the strength of structures.
Soffit	The underside of an overhanging structure.
Skewed	Slanted or not forming a straight line. Skew is the angle between a line crossing the roadway and a line normal to the roadway centerline.
SHPO	State Historic Preservation Office.
Slab	A bridge type, generally used in short structures, in which the roadway deck and its support are integral.
Span	The distance between the supports of a beam, arch, or the like.
Spandrel	The area between the exterior curves of an arch and the roadway.
Structure	In the context of this study, a term frequently substituted for "bridge".

Suspension	A bridge which suspends the roadway from high towers through a combination of cables.
Swing	A moveable bridge which pivots about a vertical axis to allow unrestricted vertical clearance of the navigation channel.
Tension members	Slender, attenuated members of a bridge which resist forces that pull them apart.
Through	Form of bridge in which traffic actually moves through the framework of a bridge.
Truss	A bridge with a framework of members, forming a triangle or system of triangles to support the weight of the bridge as well as the live or passing loads. (The nomenclature for the components of a representative through truss is shown on the following page.)
Undercrossing or underpass	A bridge structure where the principal or subject transportation facility is the lower roadway (of two roadway levels).
Vertical lift span	A moveable bridge which can be raised vertically by weights and pulleys operating in towers at each end of the structure. During raising and lowering, the bridge remains in a horizontal position.
Viaduct	Usually a bridge built over dry land or over a wide valley and consisting of a number of small spans. (Several structures in Oregon are called viaducts even though they cross waterways.)
WPA	Works Projects Administration, a federal agency (1935-1943) charged with instituting and administering public works during the Great Depression. Its original title was the Works Progress Administration.
Wrought iron	A comparatively pure form of iron, almost entirely free of carbon and having a fibrous structure that is readily forged and welded.

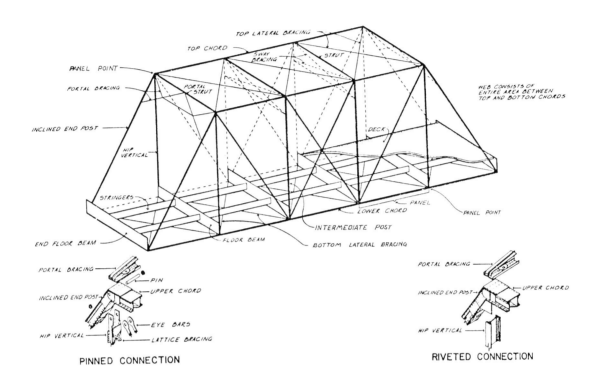

Representative truss nomenclature. (Source: Historic American Engineering Record, National Park Service, United States Department of the Interior, Washington, D.C.)

APPENDICES

St. John's Bridge (1931), Willamette River, Portland, Multnomah County

APPENDIX A

GUIDE TO THE BRIDGE PHOTO-DESCRIPTION PAGES

Bridge Photo:
The bridge photographs, with only two exceptions, are current views taken in conjunction with the historic bridges study. The exceptions are the photographs of the Yaquina Bay (Newport) Bridge, taken about 1975, and the Oswego Creek Bridge, 1980.

Bridge Name:
Bridges are frequently identified by several names. In an attempt to minimize the confusion, a standardized naming approach was used in the study. The bridge is formally named for the waterway or other feature it crosses, with appropriate other common names in parenthesis. In addition to the waterway crossed, bridges are sometimes named for their locality, geographic features, roadway or street carried, pioneer families, important persons, and so on. Viaducts and over/undercrossings usually carry the name of the roadway, but some are also named for the locality. (Note: the index at the end of the document cross-references all of the names of each bridge.)

Structure Number:
This identification is usually the number assigned to the bridge by the ODOT and is not necessarily the number used by the cities or counties. Some bridges are unnumbered, and for others, the numbers are unknown.

Roadway:
The state highway, county road, or city street carried by the bridge is shown on this line. The milepoint (M.P.) of the bridge is indicated on state highways and other highways and roads where known. The state highway numbers used in this document are not the familiar state or national route designation. The state highway numbers used in this document are shown on the following pages.

Location:
This reference indicates the city, community, or vicinity location of the bridge, followed by the county. Some bridges connect two cities, counties, and/or states, and this is so indicated. (Refer to the location information in Appendix C.)

Construction Date:
The construction date refers to the year when the bridge was fully completed and opened to service. (This date may not always agree with bridge plaques or other bridge records which show dedication dates or the major year when construction was underway.)

Material and Type:
The basic material and structural type of the main or center span are indicated in this area. Specific truss types, if appropriate, are included in parenthesis. (Refer to Appendix D.)

Ownership:
The owner is shown on this line. Unless otherwise indicated, the "State of Oregon" refers to structures under the ownership and maintenance authority of the Oregon Department of Transportation.

Narrative:
A short narrative provides descriptive information about the bridge, its history, and significance.

STATE HIGHWAY NUMBERS

The highway numbers on the photo-description pages are those used by the Oregon Department of Transportation and are not the state or federal route numbers, commonly recognized by the public. The guide below shows the highway numbers, highway names and route numbers. Highway route numbers may traverse the highways in entirety or only in part. A route common with another for less than one mile is not shown. All of the state highways are not listed, only those mentioned in this report.

PRIMARY HIGHWAYS

Highway Number	Highway Name	Route Number
1	Pacific Highway	Interstate 5, ORE99, ORE99E, ORE138, US30
1E	Pacific Highway East	ORE99E, ORE214
1W	Pacific Highway West	ORE99, ORE99W, ORE126, ORE126 Business, ORE10
2	Columbia River	Interstate 84, US30, US395, US730
2W	Lower Columbia River	US30
3	Oswego	ORE43
4	The Dalles-California	US20, US26, US97, US197, ORE140, ORE216
5	John Day	US26, US395, ORE19, ORE207
6	Old Oregon Trail	Interstate 84, US30, US395
9	Oregon Coast	US26, US101
10	Wallowa Lake	ORE82
15	McKenzie	ORE126, ORE242, ORE126 Business
16	Santiam	US20, ORE126
20	Klamath Falls-Lakeview	ORE39, ORE140
25	Redwood	ORE99, US199
26	Mt. Hood	US26, US30, ORE35
31	Albany-Corvallis	US20
33	Corvallis-Newport	US20, ORE34
35	Coos Bay-Roseburg	ORE99, ORE42
39	Salmon River	ORE18, ORE22, ORE233
45	Umpqua	ORE99, ORE38
46	Necanicum	ORE53
47	Sunset	US26, ORE47
61	Stadium	Interstate 405

SECONDARY HIGHWAYS

Highway Number	Highway Name	Route Number
105	Warrenton	US101 Business
123	Northeast Portland	US30 Bypass
125	Crown Point	—
161	Woodburn-Estacada	ORE211
164	Jefferson	—
191	Kings Valley	ORE223
210	Corvallis-Lebanon	ORE34
212	Halsey-Sweet Home	ORE228
234	Oakland-Shady	ORE99
237	Myrtle Creek	ORE99
244	Coquille-Bandon	ORE42S
260	Rogue River Loop	—
271	Sams Valley	ORE234,ORE99
272	Jacksonville	ORE238
273	Siskiyou	—
281	Hood River	—
292	Mosier-The Dalles	US30

APPENDIX B

SIGNIFICANT BRIDGE DESIGNERS

Several internationally prominent engineers have designed bridges in Oregon. Short sketches about six of these designers are below. Charles Purcell and Conde McCullough, state bridge engineers in Oregon, enjoyed reputations beyond the boundaries of the state. Ralph Modjeski, Gustav Lindenthal, David Steinman, and Joseph Strauss are luminaries in the history of bridge building in America.

Charles H. Purcell (1883-1951)

C. H. Purcell has the distinction of being the first state bridge engineer in Oregon, serving in that position from 1913-17.

Purcell was born at North Bend, Nebraska, in 1883. He received his early education there before going on to Stanford and the University of Nebraska, where he was awarded a B.S. in civil engineering in 1906. In the same year he began his professional career, serving as a resident engineer for the Union Pacific Railroad in Wyoming. During 1907-08, Purcell was a structural designer in Ely, Nevada, for the American Smelting and Refining Company. In 1909-10, he was an assistant chief engineer for the Ceno de Pasco Company, large South American copper producers, serving in New York and Peru. He returned to the Pacific Northwest in 1911 and worked for the Yuba Construction Company and the Washington Northern Railroad. In 1913, he joined the newly-formed Oregon State Highway Department.

In 1917, Purcell was appointed bridge engineer for the United States Bureau of Public Roads and in 1919 became district engineer for that bureau in Portland, Oregon. He served in that position until February 1927, when he became state highway engineer for the State of California. In 1932, because of his qualifications and notable professional record, Purcell was appointed member and secretary of the Hoover-Young Commission to locate the projected San Francisco Bay Bridge. In 1933 he was appointed chief engineer for the project. The $73,000,000 bridge was opened in November 1936 and was the largest suspension bridge in the world at that time. The Bay Bridge, with its overall 43,500-foot length, includes both suspension (West Bay Crossing) and steel cantilever (East Bay Crossing) spans.

Although the construction of the Bay Bridge tends to overshadow Purcell's other achievements, he is also famous for designing the Bixby Creek Bridge (sometimes called the Rainbow Bridge) on California's coast highway near Carmel. This classic arch span was built in 1931-33 in partnership with F.W. Panhorst. This reinforced concrete structure has unusually pure and graceful lines and is an outstanding example of Purcell's engineering abilities.

Purcell was a significant force in the design and construction of numerous bridges in Oregon during the period 1913-17 while he served as state bridge engineer. Along with his designers, K.R. Billner and L.W. Metzger, Purcell created many of the structures on the Columbia River Highway in the Columbia Gorge. (These structures are now listed on the National Register as the Columbia River Highway Historic District. They are also a National Historic Civil Engineering Landmark, as designated by the American Society of Civil Engineering.)

Conde B. McCullough (1887-1946)

Conde B. McCullough's long period of service with the State Highway Department as bridge engineer and assistant state highway engineer left a legacy of fine bridges in Oregon.

McCullough was born in Redfield, South Dakota, the son of a physician. He graduated with a B.S. in civil engineering from Iowa State College in 1910. After his graduation, his first engineering position was with the March Engineering Company, Des Moines, Iowa. From 1911 to 1916, he was employed by the Iowa State Highway Department, first as a designing engineer and then as assistant state highway engineer. During this time he designed his first bridges, the field in which he was later to achieve international recognition.

Upon moving to Oregon in 1916, McCullough joined the civil engineering department at Oregon State College (University) and was promoted to professor and head of the civil engineering department in 1918. He left teaching in 1919 to join the State Highway Department as state bridge engineer. After establishing himself at the Highway Department, McCullough also studied law at Willamette University, graduated with a Bachelor of Law degree in 1928, and was admitted to the Oregon State Bar in the same year. (He received an honorary doctorate degree in engineering from Oregon State College in 1934.) In 1932 he was appointed assistant state highway engineer and served in dual roles (also, state bridge engineer) until 1935. Between October 1935 through 1936, he was on a temporary leave, supervising the design and construction of bridges on the Inter-American Highway in Central America for the U.S. Bureau of Public Roads. He worked on the design and construction of bridges in Panama, Guatemala, El Salvador, Honduras, and Costa Rica.

Upon returning in 1937 to the Oregon State Highway Department, he served as assistant state highway engineer and spent much of that time researching and writing technical reports and books. Three technical bulletins received national recognition: "The Economics of Highway Planning," "The Determination of Highway Systems Solvencies," and "An Analysis of the Highway Tax Structure in Oregon."

McCullough was also the author of several textbooks on structural subjects. McCullough coauthored a definitive textbook on the analysis and design of elastic arch bridges. Later, he became interested in suspension bridges and was coauthor of five technical booklets on that subject. With his son, John, he wrote a two-volume work entitled *The Engineer at Law*(1946). He worked for the highway department until his death in 1946.

His bridges were noted for their beauty and innovation. Among the best known are the steel arches over the Willamette River at Oregon City and Yaquina Bay at Newport, the concrete arches on the Oregon Coast Highway, and the steel cantilever over Coos Bay. The McLoughlin Bridge across the Clackamas River south of Portland on the Pacific Highway, a series of tied steel arches with concrete approaches, was built in 1933 and was rated by the American Institute of Steel Construction as the most beautiful steel bridge in its class that year.

Louis Pierce, in his article "Esthetics in Oregon Bridges--McCullough to Date" (1980), summarizes McCullough's significance:

> McCullough was, and is, the outstanding bridge engineer in Oregon's history. In the technical sense, his research, writing, and engineering design were all first rate. His receptiveness to new ideas is evidenced by two examples that come to mind: the first tied concrete arch in America (the Wilson River Bridge, Tillamook County), and the first use of the Freyssinet method of precompression at the arch crown at Rogue River...The esthetic art of his engineering is witnessed by the beauty of the bridges that were built under his direction, and the acclaim that they have received ever since.

David Plowden, in *Bridges: Spans of North America* (1974), states that McCullough's best examples, representing perhaps the most interesting concentration of concrete bridges in America, are found on the Oregon Coast Highway on the Oregon Coast.

The art of McCullough's bridge engineering culminated in 1936 when five major bridges, all incorporating the arch form and all crossing tidal estuaries and rivers on the Pacific Ocean, were constructed on the Oregon Coast Highway. The total cost of $5,400,000 was financed through a loan and grant agreement with the federal Public Works Administration. This was by far the largest single undertaking by the bridge department up to that time and resulted in a series of magnificent structures that complement the great natural beauty of the Oregon coastline. He was honored posthumously in 1947 when the Coos Bay Bridge was renamed and dedicated the McCullough Memorial Bridge.

Many of McCullough's bridges are eligible for or have been listed on the National Register. The Rogue River (Gold Beach) Bridge was designated a Historic National Civil Engineering Landmark in 1982 by the American Society of Civil Engineers.

Ralph Modjeski (1861-1940)

David Plowden states that probably no man produced more characteristically American bridges than Ralph Modjeski. His career encompassed two eras of bridge design, beginning in the age of the steel truss and the railway and continuing into the heyday of the suspension bridge. Modjeski was a principal bridge engineer for the railroads and cities of America. Modjeski is credited with designing both railroad and highway bridges in Oregon.

Modjeski was born in Krakow, Poland, and traveled to the United States in 1878. Modjeski entered the Ecole des Ponts and Chaussees in Paris, and in 1885, after graduating at the head of his class, returned to America. After working on various assignments, Modjeski opened his own office in Chicago, launching what was to become one of the most diversified careers in all of bridge engineering. As chief engineer, he was in charge of the construction or the rebuilding of thirty of America's major bridges, four of which held records and have attained the status of classics. Although not in the latter category, the work that at one stroke elevated Modjeski to the front ranks of his profession was the building of the Thebes Bridge (1904) over the Mississippi at Thebes, Illinois. Some of Modjeski's other major bridges outside Oregon include the McKinley Bridge (1902-10) in St. Louis; the Ohio River Bridge of the Burlington Northern Railroad at Metropolis, Illinois (1914-1917); the Philadelphia-Camden Bridge (also known as Ben Franklin Bridge), Delaware River, (1922-26); and the Huey P. Long Bridge in New Orleans, (1933-36).

In Oregon, Modjeski is associated with the Willamette River (Broadway) Bridge, a steel truss with Rall-type double bascule (1913). He is also responsible for the Spokane, Portland, and Seattle (now Burlington-Northern) railroad and highway structures in North Portland, completed in 1908-09. The steel braced-spandrel arch which Modjeski designed for the Oregon Trunk Railroad was built in 1911 across the Crooked River canyon in Central Oregon and is located just downstream from C.B. McCullough's 1926 Crooked River (High) Bridge.

Gustav Lindenthal (1850-1935)

Lindenthal was born in Brunn, Austria (Czechoslovakia). He received his technical training at the Politechnicum College in Dresden and sailed for the United States in 1874. The country was then in the midst of a depression, and the only job he could get was as a stoneman on the construction of the Centennial Exposition in Philadelphia. As economic conditions improved, Lindenthal became a designer for the Keystone Bridge Company in Pittsburgh, and in 1878 was offered the post of bridge engineer for the Atlantic-Great Western Railroad in Cleveland. Three years later he returned to Pittsburgh, and shortly after setting up a private practice there, secured the contract for the Smithfield Street Bridge. This large and successful lenticular truss bridge across the Monongahela River in Pittsburgh took the place of a suspension bridge built in 1845 by John A. Roebling. Lindenthal's major claims to fame are three bridges on the East River in New York--Blackwell's Island or Queensboro Bridge (1908), steel cantilever truss; Manhattan Bridge (1908), suspension; and the Hell Gate Arch (1917). The latter bridge was a two-hinged braced spandrel-arch, and is still the world's highest arch bridge.

Before the Hell Gate Arch was opened, Lindenthal was already at work on a structure that was to equal Hell Gate in every respect--the Chesapeake and Ohio Railroad's Sciotoville Bridge (1914), a continuous truss across the Ohio River near Portsmouth, Ohio. Lindenthal's decision to select this bridge form represented a much more daring solution than the one adopted at the East River.

His last commission came from the City of Portland for a series of bridges across the Willamette River. The Sellwood and Ross Island bridges were completed in 1925 and 1926, respectively, when Lindenthal was in his mid-70s. Lindenthal also took over the final design and construction of the Burnside Bridge (1926).

David B. Steinman (1886-1960)

David Steinman was born in New York City and attended the City College of New York and Columbia University. He received a doctorate from Columbia in 1911. He was professor of civil engineering at the University of Idaho and professor of civil and mechanical engineering at the City College of New York.

Steinman's bridge-building career began as an assistant to Gustav Lindenthal between 1914 and 1917, assisting on the Hell Gate Arch Bridge (1917) and the Sciotoville Bridge (1914). From the 1930s onward, Steinman, along with Othmar H. Amman (later the designer of the George Washington and Verrazano-Narrows bridges in New York City), dominated the American bridge-building scene. Steinman was a prolific builder, with many of his structures being record- breaking suspension bridges. Famous bridges by Steinman include the Henry Hudson Bridge (1936), the longest hingeless arch and the longest plate girder arch span in the world; the Thousand Islands International Bridge (1938) over the St. Lawrence River; the Carquinez Strait Bridge (1927), California; and Mackinac Strait Bridge, Mackinaw City-St. Ignace, Michigan (1957), steel suspension, 3,800 feet. Steinman also designed or acted as consulting engineer on several bridges outside the United States.

Steinman's contribution to Oregon is the St. John's suspension bridge across the Willamette River in North Portland, completed in 1931. This is Steinman's only constructed design in Oregon, although he did work on designs between 1929 and 1935 for the Astoria Bridge across the Columbia River.

He authored *The Builders of the Bridge* (1945), several standard works on bridge design and construction, and numerous technical and interpretive articles.

Joseph B. Strauss (1870-1938)

Joseph Strauss' place in engineering history became solidly fixed in 1937 with the completion of the Golden Gate Suspension Bridge at San Francisco. He was the chief engineer and gained world fame for what David Plowden calls "...one of the world's greatest engineering achievements, an outstanding masterpiece of modern American bridge design."

Strauss graduated from the University of Cincinnati in 1882, where for his thesis he designed a span over the Bering Straits between Alaska and Russia. Two years later he started his own engineering firm. Prior to his association with the Golden Gate, his name was best known as a designer of bascule bridges. (The Burnside Bridge across the Willamette River in Portland includes a bascule system designed by Strauss.) His most famous bascule span is the Arlington Memorial Bridge in Washington, D.C.

Strauss also completed several large steel cantilever bridges. The Lewis and Clark Bridge (1930) across the Columbia River, with its 1,200-foot main span, was the longest cantilever span in the United States when completed. Plowden considers the bridge one of the most beautiful of all cantilevers.

Strauss completed bridges in Europe, Russia, Japan, Egypt, China, and South America. In total, he designed nearly four hundred bridges.

BRIDGE CONTRACTORS

Many of the bridge contractors active in Oregon from the turn of the century to the late 1930s are listed below. The names of bridge contractors are unknown for many of the bridges constructed during this period. This list indicates the contractors associated with the pre-1941 bridges. The names were obtained primarily from bridge nameplates, bridge plans and contracts, and articles in engineering journals and other publications, including the Oregon State Highway Commission's biennial reports.

BRIDGE CONTRACTORS

Name	Number of Bridges	Period
American Bridge Company of New York	1	1913
Albert Anderson	1	1917
John W. Ash, Builder	1	1916
J.J. Badraun	3	1926-34
E.F. Balgeman	1	1930
Booth and Pomeroy	1	1926
Bullen Bridge Company	2	1894
James J. Burke Company	1	1922
C.A. Catching	1	1931
Civilian Conservation Corps	1	1936
Clackamas Construction	9	1926-31
Coast Bridge Company	5	1911-14
Columbia Bridge Company	4	1909-10
Colonial Building Company	1	1920
William D. Comer and Wesley Vandercook	1	1929
The Construction Company	4	1914
H.E. Doering	2	1923-27
Dolan Construction Company	1	1935
Clifford A. Dunn	1	1938
D.T. Eaton Construction Company	1	1928
P.S. Easterday and Company	1	1906-07
A.B. Gidley	1	1922
Gilpin Construction Company (and General Construction Company)	2	1925-36
Gray and Chandler, Engineers	1	1924
A. Guthrie and Company	1	1922
Hargreaves and Lindsay	2	1931
J.S. and J.R. Hillstrom	2	1934-39
John K. Holt	2	1931
Hurley-Mason and Company	1	1913
Illinois Steel Bridge Company	1	1921
International Contract Company	2	1907
J.F. Johnson	1	1935
Joplin and Eldon	1	1934
Kelly and Lilly (with the American Bridge Company)	1	1921
A.D. Kern	2	1918-20
Kern and Kibbe	1	1936

A.H. Kingsbury	1	1929
Rudolf K. Krausse	1	1929
Edward Kreig	1	1930
Kuckenberg-Wittman	4	1926-29
Mercer-Fraser Company	2	1932-36
L.W. Metzger	1	1925
C.J. Montag and Sons	5	1926-39
Morrison-Knudsen	1	1925
Mountain States Construction Company	5	1934-39
J. Elmer Nelson	1	1921
Northwest Contract Company	1	1929
F.L. Odom	2	1929
F.L. Odom and P.L. Frazier	2	1931
Olds Construction Company	1	1926
Pacific Bridge Company	6	1914-26
Pacific Irons Work	1	1905
Parker and Banfield	3	1919-36
Pearson Construction Company	1	1917
Penn Bridge Company	1	1910
Pennsylvania Steel Company	2	1910-13
O.N. Pierce Company	5	1924-32
D.P. Plymale	1	1927
J.H. Pomeroy Company and John A. Roebling's Sons Company	1	1931
Poole and McGonigle	1	1928
Portland Bridge Company	6	1914-36
Porter Brothers (with American Company et. al.)	1	1917
George G. Reeves	1	1925
Rigdon Brothers	1	1929
Joe Rocco	1	1926
Settegren Brothers	1	1934
Sig Ash	1	1937
J.W. Sweeney	2	1914
Teufel and Carlson	1	1936
Tobin and Pierce	2	1922
Union Bridge Company	8	1922-31
J.S. Varner	1	1929
Virginia Bridge and Iron Company	2	1930-36
Waddell and Harrington	4	1910-12
Robert Wakefield	1	1912
Wauna Toll Bridge Company	1	1926
Charles O. Young	1	1931
TOTAL	149	

APPENDIX C

BRIDGE STATISTICS

This appendix provides some basic statistics about the historic bridges including distribution, age by type, and ownership. The bridges are also listed by county and alphabetically by the name of the city or area within the county. In the location list, the historic highway bridges (National Register eligible or listed) are shown in bold type, and the reserve and notable post-1940 constructed bridges (presently considered National Register ineligible) are in regular type. Photo-descriptions of the reserve and notable post-1940 bridges are in Appendices E and F, respectively.

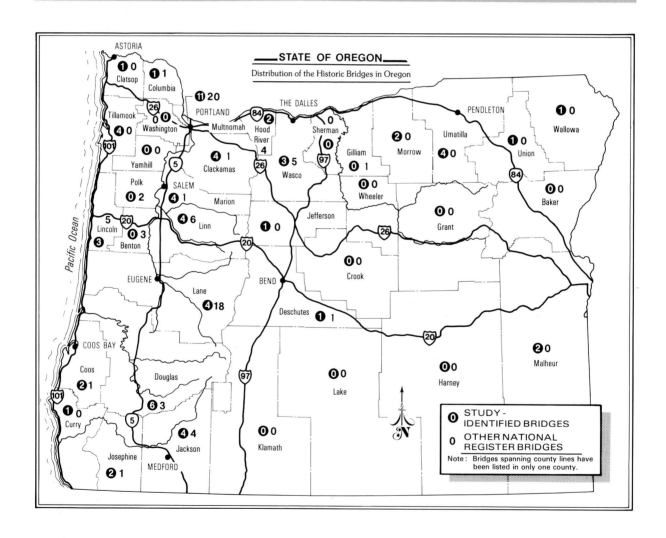

TYPE AND PERIOD OF CONSTRUCTION OF THE HISTORIC BRIDGES

	Pre-1900	1900-09	1910-19	1920-29	1930-40	Post-1940	Total
Metal Truss	2	8	8	5	2	-	25 (17%)
Timber Truss	-	-	7	16	18	6	47 (33%)
Timber Arch	-	-	-	-	1	-	1 (0.5%)
Reinforced Concrete Arch	-	-	10	15	6	-	31 (21%)
Steel Arch	-	-	-	1	2	-	3 (2%)
Moveable	-	-	4	2	2	-	8 (6%)
Suspension	-	-	-	-	1	-	1 (0.5%)
Slab, Beam, and Girder	-	-	17	8	2	2	29 (20%)
TOTAL	2 (1%)	8 (6%)	46 (32%)	47 (32%)	34 (23%)	8 (6%)	145 (100%)
All Pre-1941 Bridges	2 (0.1%)	13 (1.1%)	108 (9%)	398 (33%)	661 (56%)	-	1182 (100%)

OWNERSHIP OF THE HISTORIC BRIDGES

	Study Identified	Old Columbia River	Covered	Other	Total
OREGON DEPARTMENT OF TRANSPORTATION	32	20	-	3	55 (38%)
COUNTY	21	4	37	1	63 (43%)
CITY	11*	-	-	1	12 (8%)
OTHER PUBLIC JURISDICTIONS	2	2	-	2	6 (5%)
PRIVATE	2	-	5	2	9 (6%)
Total	68 (47%)	26 (18%)	42 (29%)	9 (6%)	145 (100%)

*One bridge is jointly-owned by the city and county, but is assigned to the city herein.

BRIDGE LOCATIONS BY COUNTY

BAKER COUNTY

Baker, Powder River (Myrtle Street) Bridge
Halfway vicinity, Pine Creek (Crow Lane) Bridge

BENTON COUNTY

Alder vicinity, Marys River Bridge
Alsea vicinity, Alsea River (Hayden) Covered Bridge
Corvallis, Willamette River (Van Buren Street) Bridge, also Linn County
Monroe vicinity, Willamette Slough (Irish Bend) Covered Bridge
North Albany, Willamette River (Albany) Bridge, also Linn County
Wren vicinity, Mary's River (Harris) Covered Bridge

CLACKAMAS COUNTY

Bull Run, Bull Run River Bridge
Estacada, Clackamas River (Estacada) Bridge
Gladstone, Clackamas River (McLoughlin) Bridge
Gladstone, Clackamas River (Park Place) Bridge
Lake Oswego, Oswego Creek Bridge
Oregon City-West Linn, Willamette River (Oregon City) Bridge
Sandy vicinity, Sandy River (Lusted Road) Bridge

CLATSOP COUNTY

Astoria, Columbia River (Astoria) Bridge, also Megler vicinity, Pacific County, Washington
Astoria vicinity, Beltline Overcrossing
Astoria vicinity, Lewis and Clark River Bridge
Astoria vicinity, Old Young's Bay Bridge
Elsie vicinity, Nehalem River Bridge
Necanicum Junction vicinity, Soapstone Creek Bridge
Seaside, Necanicum River (Seaside) Bridge

COLUMBIA COUNTY

Clatskanie vicinity, Beaver Creek (Kukkla Road) Bridge
Rainier, Columbia River (Lewis and Clark) Bridge, also Longview, Cowlitz County, Washington
St. Helens, Lower Milton Creek (McDonald) Bridge

COOS COUNTY

Coquille, Coquille River Bridge
North Bend, Coos Bay (McCullough Memorial) Bridge
Remote, Sandy Creek (Remote) Covered Bridge

CROOK COUNTY

Prineville vicinity, Crooked River (Elliott Lane) Bridge

CURRY COUNTY

Brookings vicinity, Thomas Creek Bridge
Ophir, Euchre Creek Bridge
Wedderburn-Gold Beach, Rogue River (Gold Beach) Bridge

DESCHUTES COUNTY

Bend vicinity, Swalley Canal (Rock O' the Range) Covered Bridge
Tumalo vicinity, Tumalo Irrigation Ditch Bridge

DOUGLAS COUNTY

Days Creek, South Umpqua River (Worthington) Bridge
Days Creek vicinity, South Umpqua River (Milo Academy) Covered Bridge
Drain, Pass Creek Covered Bridge
Drain vicinity, Elk Creek (Roaring Camp) Covered Bridge
Elkton vicinity, Elk Creek (First Crossing) Bridge
Elkton vicinity, Elk Creek (Second Crossing) Bridge
Elkton vicinity, Elk Creek (Third Crossing) Bridge
Elkton vicinity, Elk Creek (Fourth Crossing) Bridge
Glide vicinity, Little River (Cavitt Creek) Covered Bridge
Myrtle Creek, South Umpqua River (Myrtle Creek) Bridge
Myrtle Creek vicinity, South Myrtle Creek (Neal Lane) Covered Bridge
Oakland, Calapooya Creek (Oakland) Bridge
Reedsport, Umpqua River (Reedsport) Bridge
Scottsburg, Umpqua River (Scottsburg) Bridge
Steamboat vicinity, North Umpqua River (Mott) Bridge
Sutherlin vicinity, Calapooya Creek (Rochester) Covered Bridge
Winchester, North Umpqua River (Winchester) Bridge
Winston vicinity, South Umpqua River (Winston) Bridge

GILLIAM COUNTY

Olex vicinity, Rock Creek (Olex) Bridge

GRANT COUNTY

None.

HARNEY COUNTY

None.

HOOD RIVER COUNTY

Cascade Locks, Columbia River (Bridge of the Gods), also Stevenson vicinity, Skamania County, Washington
Cascade Locks vicinity, Ruckel Creek Bridge
Hood River, Columbia River (White Salmon) Bridge, also White Salmon, Klickitat County, Washington
Hood River vicinity, Hood River (Tucker) Bridge
Hood River vicinity, Rock Slide Viaduct
Hood River vicinity, Ruthton Point Viaduct
Mount Hood vicinity, South Fork Hood River (Sahalie Falls) Bridge
Wyeth vicinity, Gorton Creek Bridge

JACKSON COUNTY

Gold Hill, Rogue River (Gold Hill) Bridge
Gold Hill vicinity, Rogue River (Rock Point) Bridge
Lakecreek vicinity, Lost Creek Covered Bridge
Medford vicinity, Antelope Creek Covered Bridge
Ruch vicinity, Applegate River (McKee) Covered Bridge
Steinman, Steinman Overcrossing
Steinman vicinity, Dollarhide Overcrossing
Wimer, Evans Creek (Wimer) Covered Bridge

JEFFERSON COUNTY

Madras vicinity, Crooked River (Lake Billy Chinook) Bridge
Terrebonne vicinity, Crooked River (High) Bridge

JOSEPHINE COUNTY

Grants Pass, Rogue River (Caveman) Bridge
Merlin vicinity, Rogue River (Robertson) Bridge
Provolt vicinity, Williams Creek Bridge
Sunny Valley vicinity, Grave Creek Covered Bridge

KLAMATH COUNTY

Klamath Falls, Link River Bridge

LAKE COUNTY

None

LANE COUNTY

Cottage Grove vicinity, Mosby Creek Covered Bridge
Cottage Grove vicinity, Row River (Currin) Covered Bridge
Crow vicinity, Coyote Creek (Battle Creek) Covered Bridge
Dexter vicinity, Lost Creek (Parvin) Covered Bridge
Dorena vicinity, Row River (Dorena) Covered Bridge
Florence, Siuslaw River (Florence) Bridge
Greenleaf vicinity, Lake Creek (Nelson Mountain) Covered Bridge
Heceta Head, Cape Creek Bridge
Heceta Head vicinity, Big Creek Bridge
Jasper vicinity, Fall Creek (Pengra) Covered Bridge
Lowell vicinity, Fall Creek (Unity) Covered Bridge
Lowell vicinity, Middle Fork Willamette River (Lowell) Covered Bridge
Marcola vicinity, Mill Creek (Wendling) Covered Bridge
Marcola vicinity, Mohawk River (Ernest) Covered Bridge
McKenzie Bridge vicinity, Horse Creek Covered Bridge
Rainbow, McKenzie River (Belknap) Covered Bridge
Richardson, Siuslaw River (Richardson) Bridge
Richardson vicinity, Wildcat Creek Covered Bridge
Springfield, Willamette River (Springfield) Bridge
Swisshome vicinity, Deadwood Creek Covered Bridge
Vida vicinity, McKenzie River (Goodpasture) Covered Bridge
Walden vicinity, Mosby Creek (Stewart) Covered Bridge

Westfir, North Fork of the Middle Fork Willamette River (Office) Covered Bridge
Yachats vicinity, Cummins Creek Bridge
Yachats vicinity, Tenmile Creek Bridge

LINCOLN COUNTY

Chitwood, Yaquina River (Chitwood) Covered Bridge
Depoe Bay, Depoe Bay Bridge
Fisher, Five Rivers (Fisher School) Covered Bridge
Lincoln City vicinity, Drift Creek Covered Bridge
Newport, Yaquina Bay (Newport) Bridge
Otter Crest vicinity, Rocky Creek (Ben Jones) Bridge
Rose Lodge vicinity, Salmon River Bridge
Waldport, Alsea Bay (Waldport) Bridge
Yachats vicinity, North Fork Yachats River Covered Bridge

LINN COUNTY

Albany, Willamette River (Albany) Bridge, also North Albany, Benton County
Cascadia, Santiam River (Cascadia Park) Bridge
Cascadia, South Fork Santiam River (Short) Covered Bridge
Corvallis vicinity, Willamette River (Van Buren Street) Bridge, also Corvallis, Benton County
Crabtree vicinity, Crabtree Creek (Hoffman) Covered Bridge
Crabtree vicinity, Crabtree Creek (Larwood) Covered Bridge
Crabtree vicinity, Thomas Creek (Weddle) Covered Bridge
Crawfordsville, Calapooia River (Crawfordsville) Covered Bridge
Jefferson vicinity, Santiam River bridges, also Marion County
Jefferson vicinity, Santiam River (Jefferson) Bridge, also Jefferson, Marion County
Lyons vicinity, Thomas Creek (Jordan) Covered Bridge
Scio vicinity, Crabtree Creek (Bohemian Hall) Covered Bridge
Scio vicinity, Thomas Creek (Gilkey) Covered Bridge
Scio vicinity, Thomas Creek (Hannah) Covered Bridge
Scio vicinity, Thomas Creek (Shimanek) Covered Bridge

MALHEUR COUNTY

Danner vicinity, Cow Creek Bridge
Jordan Valley vicinity, Jordon Creek Overflow Bridge
Rome vicinity, Owyhee River Bridge

MARION COUNTY

Jefferson, Santiam River (Jefferson) Bridge, also Jefferson vicinity, Linn County
Jefferson vicinity, Santiam River bridges, also Linn County
Salem, Mill Creek (Front Street, N.E.) Bridge
Salem, Mill Creek (Summer Street, N.E.) Bridge
Salem, Pringle Creek (Commercial Street, South) Bridge
Salem, Pringle Creek (Liberty Street, S.E.) Bridge
Salem, Pringle Creek/Shelton Creek (Church Street, S.E.) Bridge
Salem, Portland Road, N.E., Undercrossing
Silverton vicinity, Abiqua Creek (Gallon House) Covered Bridge

MORROW COUNTY

Cecil, Willow Creek (Cecil) Bridge

Jordan vicinity, Rhea Creek Bridge
Ruggs vicinity, Rhea Creek (Spring Hollow) Bridge

MULTNOMAH COUNTY

Bonneville vicinity, Eagle Creek Bridge
Bonneville vicinity, Moffett Creek Bridge
Bonneville vicinity, Tanner Creek Bridge
Bonneville vicinity, Toothrock and Eagle Creek viaducts
Bridal Veil vicinity, Bridal Veil Falls Bridge
Crown Point, Crown Point Viaduct
Gresham vicinity, Sandy River (Stark Street) Bridge
Gresham vicinity, Stark Street Viaduct
Latourell, Latourell Creek Bridge
Latourell vicinity, Young Creek (Shepperd's Dell) Bridge
Multnomah Falls, Multnomah Creek Bridge
Multnomah Falls vicinity, East Multnomah Falls Viaduct
Multnomah Falls vicinity, Wahkeena Falls Bridge
Multnomah Falls vicinity, West Multnomah Falls Viaduct
Oneonta, Oneonta Gorge Creek (Old) Bridge
Oneonta, Oneonta Gorge Creek (New) Bridge
Oneonta, Horsetail Falls Bridge
Portland, Balch Gulch Bridge
Portland, Columbia Slough (N.E. Union Avenue) Bridge
Portland, Johnson Creek Bridge
Portland, N.E. 12th Avenue Overcrossing
Portland, N.E. Grand Avenue Overcrossing
Portland, North Fessenden Street Overcrossing
Portland, North Lombard State Overcrossing
Portland, North Willamette Boulevard Overcrossing
Portland, N.W. Alexandra Avenue Viaduct
Portland, S.W. Vista Avenue Viaduct
Portland, Willamette River (Broadway) Bridge
Portland, Willamette River (Burnside) Bridge
Portland, Willamette River (Fremont) Bridge
Portland, Willamette River (Hawthorne) Bridge
Portland, Willamette River (Ross Island) Bridge
Portland, Willamette River (Steel) Bridge
Portland, Willamette River (St. John's) Bridge
Portland vicinity, Columbia River (Interstate Northbound) Bridge, also Vancouver, Clark County, Washington
Portland vicinity, McCarthy Creek Bridge
Troutdale, Beaver Creek (Sandy River Overflow) Bridge
Troutdale, Sandy River (Troutdale) Bridge

POLK COUNTY

Dallas vicinity, Rickreall Creek (Pumping Station) Covered Bridge
Kings Valley vicinity, Luckiamute River (Grant Road) Bridge
Kings Valley vicinity, Ritner Creek Covered Bridge
Willamina, South Yamhill River (Steel) Bridge

SHERMAN COUNTY

None

TILLAMOOK COUNTY

Arch Cape vicinity, Necarney Creek Bridge
Manzanita vicinity, Chasm (Neahkahnie Mountain) Bridge
Tillamook vicinity, Killam Creek Bridge
Tillamook vicinity, Wilson River Bridge

UMATILLA COUNTY

Adams vicinity, Greasewood Creek Bridge
Pendleton, Old Mill Race Bridge
Pendleton, Umatilla River (S.E. 8th Street) Bridge
Pendleton, Umatilla River (S.W. 10th Street) Bridge
Umapine vicinity, Pine Creek Bridge, No. 59C528
Umapine vicinity, Pine Creek Bridge, No. 59C534
Umatilla, Umatilla River (Umatilla) Bridge
Weston vicinity, Pine Creek Bridge, No. 10712

UNION COUNTY

Elgin vicinity, Grande Ronde River (Old Rhinehart) Bridge
Imbler vicinity, Grande Ronde River (McKennon Road) Bridge
La Grande vicinity, Grande Ronde River (Perry Overcrossing) Bridge
Palmer Junction, Grande Ronde River (Yarrington) Bridge

WALLOWA COUNTY

Troy, Grande Ronde River (Troy) Bridge

WASCO COUNTY

Boyd vicinity, Fifteenmile Creek (Adkisson) Bridge
Maupin, Deschutes River (Maupin) Bridge
Mosier, Mosier Creek Bridge
Mosier, Rock Creek Bridge
Mosier vicinity, Mosier Creek (State Road) Bridge
Rowena vicinity, Dry Canyon Creek Bridge
Rowena vicinity, Hog Creek Canyon (Rowena Dell) Bridge
The Dalles, Mill Creek (West Sixth Street) Bridge
The Dalles vicinity, Chenoweth Creek Bridge
The Dalles vicinity, Fifteenmile Creek (Seufert) Viaduct

WASHINGTON COUNTY

None.

WHEELER COUNTY

None.

YAMHILL COUNTY

McMinnville vicinity, North Yamhill River Bridge.

APPENDIX D

BRIDGES BY STRUCTURAL TYPE

The list classifies the bridges shown in this document by the structural type of the main or center span. Specific types follow the five general types-truss, arch, suspension, moveable, and slab, beam and girder. Within each specific type, the bridges are arranged chronologically by date of construction and then alphabetically (if several bridges were constructed in the same year). The historic highway bridges, those determined eligible for or listed on the National Register, are printed in bold type. The reserve bridges and notable post-1940 constructed bridges, not considered National Register eligible, are shown in regular type.

Truss

TIMBER DECK TRUSS (HOWE)

Santiam River (Cascadia Park) Bridge, Linn County, 1928.

TIMBER THROUGH TRUSS (HOWE)

Drift Creek Covered Bridge,* Lincoln County, 1914.
Rickreall Creek (Pumping Station) Covered Bridge, Polk County, 1916.
Abiqua Creek (Gallon House) Covered Bridge, Marion County, 1917.
Applegate River (McKee) Covered Bridge, Jackson County, 1917.
Alsea River (Hayden) Covered Bridge, Benton County, 1918.
Five Rivers (Fisher School) Covered Bridge, Lincoln County, 1919.
Grave Creek Covered Bridge, Lane County, 1920.
Mosby Creek Covered Bridge, Lane County, 1920.
Lost Creek (Parvin) Covered Bridge, Lane County, 1921.
Sandy River (Remote) Covered Bridge, Coos County, 1921.
Antelope Creek Covered Bridge, Jackson County, 1922.
Coyote Creek (Battle Creek) Covered Bridge, Lane County, 1922.
Pass Creek Covered Bridge, Douglas County, 1925.
Row River (Currin) Covered Bridge, Lane County, 1925.
Wildcat Creek Covered Bridge, Polk County, 1925.
Yaquina River (Chitwood) Covered Bridge, Lincoln County, 1926.
Ritner Creek Covered Bridge, Polk County, 1927.
Lake Creek (Nelson Mountain) Covered Bridge, Lane County, 1928.
Elk Creek (Roaring Camp) Covered Bridge, Douglas County, 1929.
Horse Creek Covered Bridge, Lane County, 1930.
Mosby Creek (Stewart) Covered Bridge, Lane County, 1930.
Calapooia River (Crawfordsville) Covered Bridge, Linn County, 1932.
Deadwood Creek Covered Bridge, Lane County, 1932.
Calapooya Creek (Rochester) Covered Bridge, Douglas County, 1933.
Crabtree (Hoffman) Covered Bridge, Linn County, 1936.
Fall Creek (Unity) Covered Bridge, Lane County, 1936.
Marys River (Harris) Covered Bridge, Benton County, 1936.
Thomas Creek (Hannah) Covered Bridge, Linn County, 1936.
Thomas Creek (Jordan) Covered Bridge, Linn County, 1937.
Thomas Creek (Weddle) Covered Bridge, Linn County, 1937.

---

* The 51 covered bridges in the document are cumulatively contained in these types: Timber Through Truss (Howe), Timber Through Truss (Kingpost), Timber Through Truss (Queenpost), Timber Deck Girder, and Steel Through Girder.

Fall Creek (Pengra) Covered Bridge, Lane County, 1938.
McKenzie River (Goodpasture) Covered Bridge, Lane County, 1938.
Mill Creek (Wendling) Covered Bridge, Lane County, 1938.
Mohawk River (Ernest) Covered Bridge, Lane County, 1938.
Crabtree Creek (Larwood) Covered Bridge, Linn County, 1939.
Thomas Creek (Gilkey) Covered Bridge, Linn County, 1939.
Little River (Cavitt Creek) Covered Bridge, Douglas County, 1943.
North Fork of the Middle Fork Willamette River (Office) Covered Bridge, Lane County, 1944.
Middle Fork Willamette River (Lowell) Covered Bridge, Lane County, 1945.
South Fork Santiam River (Short) Covered Bridge, Linn County, 1945.
Crabtree Creek (Bohemian Hall) Covered Bridge, Linn County, 1947.
Row River (Dorena) Covered Bridge, Lane County, 1949.
Willamette Slough (Irish Bend) Covered Bridge, Benton County, 1954.
McKenzie River (Belknap) Covered Bridge, Lane County, 1966.
Thomas Creek (Shimanek) Covered Bridge, Linn County, 1966.

TIMBER THROUGH TRUSS (KINGPOST)

South Myrtle Creek (Neal Lane) Covered Bridge, Douglas County, 1939.

TIMBER THROUGH TRUSS (QUEENPOST)

Lost Creek Covered Bridge, Jackson County, 1919.
Evans Creek (Wimer) Covered Bridge, Jackson County, 1927.
North Fork Yachats River Covered Bridge, Lincoln County, 1938.

IRON THROUGH TRUSS (PENNSYLVANIA-PETIT)

Bull Run River Bridge, Clackamas County, 1894.
Sandy River (Lusted Road) Bridge, Clackamas County, 1894.

STEEL TRUSS (CANTILEVER)

Willamette River (Ross Island) Bridge, Multnomah County, Deck Truss, 1926.
Columbia River (Bridge of the Gods), Hood River County (Oregon) and Skamania County (Washington), Through Truss, 1926.
Columbia River (Lewis and Clark) Bridge, Columbia County (Oregon) and Cowlitz County (Washington), Through Truss, 1930.
Coos Bay (McCullough Memorial) Bridge, Coos County, Through Truss, 1936.
Columbia River (Astoria) Bridge, Clatsop County (Oregon) and Pacific County (Washington), Through Truss, 1966.

STEEL THROUGH TRUSS (PARKER)

Owyhee River Bridge, Malheur County, 1906.
Rogue River (Robertson) Bridge, Josephine County, Ca. 1909.
Sandy River (Stark Street) Bridge, Multnomah County, 1914.
Clackamas River (Park Place) Bridge, Clackamas County, 1921.
Willamette River (Albany) Bridge, Linn-Benton counties, 1925.

STEEL THROUGH TRUSS (PENNSYLVANIA-PETIT)

Grande Ronde River (Troy) Bridge, Wallowa County, 1910.
South Umpqua River (Worthington) Bridge, Douglas County, Ca. 1910.

STEEL DECK TRUSS (PRATT)

Balch Gulch Bridge, Multnomah County, 1905.

STEEL PONY TRUSS (PRATT AND VARIATIONS)

Powder River (Myrtle Street) Bridge, Baker County, 1920.
Rock Creek (Olex) Bridge, Gilliam County, Half-Hip Pratt, Ca. 1905.
Willow Creek (Cecil) Bridge, Morrow County, Half-Hip Pratt, 1909.

STEEL THROUGH TRUSS (PRATT AND VARIATIONS)

Grande Ronde River (Yarrington) Bridge, Union County, 1906.
Rhea Creek (Spring Hollow) Bridge, Morrow County, 1909.
Umatilla River (S.E. 8th Street) Bridge, Umatilla County, 1909.
Cow Creek Bridge, Malheur County, Ca. 1910.
South Yamhill River (Steel) Bridge, Polk County, Ca, 1910.
Sandy River (Troutdale) Bridge, Multnomah County, 1912.
Luckiamute River (Grant Road) Bridge, Polk County, 1918.

STEEL PONY TRUSS (QUEENPOST)

Jordon Creek Overflow Bridge, Malheur County, Ca. 1910.
Pine Creek Bridge, No. 10712, Umatilla County, Ca. 1910.
Pine Creek Bridge, No. 59C528, Umatilla County, Ca. 1910.
Pine Creek Bridge, No. 59C534, Umatilla County, Ca. 1910.

STEEL DECK TRUSS (WARREN AND VARIATIONS)

North Fessenden Street Overcrossing, Multnomah County, 1909.
North Lombard Street Overcrossing, Multnomah County, 1909.
North Willamette Street Overcrossing, Multnomah County, 1909.
North Yamhill River Bridge, Yamhill County, Warren with Verticals, 1921.
Grande Ronde River (Old Rhinehart) Bridge, Union County, Warren with Verticals, 1922.
Calapooya Creek (Oakland) Bridge, Douglas County, 1925.
Deschutes River (Maupin) Bridge, Wasco County, 1929.
Elk Creek (First Crossing) Bridge, Douglas County, Warren with Verticals, 1931.
Elk Creek (Third Crossing) Bridge, Douglas County, Warren with Verticals, 1931.
Elk Creek (Fourth Crossing) Bridge, Douglas County, Warren with Verticals, 1931.
Thomas Creek Bridge, Curry County, Warren with Verticals, 1961.

STEEL PONY TRUSS (WARREN AND VARIATIONS)

Greasewood Creek Bridge, Umatilla County, 1907.
Pine Creek (Crow Lane) Bridge, Baker County, 1911.
Umatilla River (S.W. 10th Street) Bridge, Umatilla County, Warren with Polygonal Upper Chords, 1913.
Marys River Bridge, Benton County, Warren with Polygonal Upper Chords, 1914.
Lower Milton Creek (McDonald) Bridge, Columbia County, 1914.
Beaver Creek (Kukkla Road) Bridge, Columbia County, Double-Intersection Warren, Ca. 1915.
Grande Ronde River (McKennon Road) Bridge, Union County, Warren with Polygonal Upper Chords, 1925.

STEEL THROUGH TRUSS (WARREN VARIATIONS)

Siuslaw River (Richardson) Bridge, Lane County, Double-Intersection Warren, 1912.
Crooked River (Elliott Lane) Bridge, Crook County, Double-Intersection Warren, Ca. 1914.
Umpqua River (Scottsburg) Bridge, Douglas County, Warren with Polygonal Upper Chords, 1929.
Willamette River (Springfield) Bridge, Lane County, Warren with Polygonal Upper Chords, 1929.

Arch

TIMBER DECK ARCH

North Umpqua River (Mott) Bridge, Douglas County, 1936.

REINFORCED CONCRETE DECK ARCH

Mill Creek (Front Street, N.E.) Bridge, Marion County, Single Span, 1913.
Killam Creek Bridge, Tillamook County, Single Span, 1914.
Latourell Creek Bridge, Multnomah County, Three Spans, 1914.
Multnomah Creek Bridge, Multnomah County, Single Span, 1914.
Tumalo Irrigation Ditch Bridge, Deschutes County, Five Spans, 1914.
Young Creek (Shepperd's Dell) Bridge, Multnomah County, Single Span, 1914.
Eagle Creek Bridge, Multnomah County, Single Span, 1915.
Moffett Creek Bridge, Multnomah County, Single Span, 1915.
Stark Street Viaduct, Multnomah County, Single Span, 1915.
Williams Creek Bridge, Josephine County, Single Span, 1917.
Mosier Creek Bridge, Wasco County, Single Span, 1920.
Oswego Creek Bridge, Clackamas County, Single Span, 1920.
Rogue River (Rock Point) Bridge, Jackson County, Single Span, 1920.
Dry Canyon Creek Bridge, Wasco County, Single Span, 1921.
N.W. Alexandra Avenue Viaduct, Multnomah County, Single Span, 1922.
South Umpqua River (Myrtle Creek) Bridge, Douglas County, Six Spans, 1922.
Grande Ronde River (Perry Overcrossing) Bridge, Union County, Six Spans, 1924.
Necanicum River (Seaside) Bridge, Clatsop County, three spans, 1924.
North Umpqua River (Winchester) Bridge, Douglas County, Seven Spans, 1924.
Fifteenmile Creek (Adkisson) Bridge, Wasco County, Single Span, 1925.
Umatilla River (Umatilla) Bridge, Umatilla County, Three Spans, 1925.
Depoe Bay Bridge, Lincoln County, Single Span, 1927.
Rocky Creek (Ben Jones) Bridge, Lincoln County, Single Span, 1927.
Rogue River (Gold Hill) Bridge, Jackson County, Single Span, 1927.
S.W. Vista Avenue Viaduct, Multnomah County, Single Span, 1927.
Soapstone Creek Bridge, Clatsop County, Single Span, 1928.
South Fork Hood River (Sahalie Falls) Bridge, Hood River County, Single Span, 1928.
Salmon River Bridge, Lincoln County, Single Span, 1930.
Cape Creek Bridge, Lincoln County, Single Span, 1931.
Cummins Creek Bridge, Lane County, Single Span, 1931.
Rogue River (Gold Beach) Bridge, Curry County, Seven Spans, 1931.
Hood River (Tucker) Bridge, Hood River County, Single Span, 1932.
Clackamas River (Estacada) Bridge, Clackamas County, Single Span, 1936.
Nehalem River Bridge, Clatsop County, Single Span, 1939.

REINFORCED CONCRETE HALF-THROUGH ARCH

Rogue River (Caveman) Bridge, Josephine County, Three Spans, 1931.

REINFORCED CONCRETE THROUGH ARCH

Santiam River (Jefferson) Bridge, Marion-Linn counties, Three Spans, 1933.

REINFORCED CONCRETE THROUGH TIED ARCH

Big Creek Bridge, Lane County, Single Span, 1931.
Tenmile Creek Bridge, Lane County, Single Span, 1931.
Wilson River Bridge, Tillamook County, Single Span, 1931.
Alsea Bay (Waldport) Bridge, Lincoln County, Three Central Spans, 1936.

STEEL DECK ARCH

Crooked River (High) Bridge, Jefferson County, Single Span, 1926.

STEEL HALF-THROUGH ARCH

Willamette River (Oregon City) Bridge, Clackamas County, Single Span, 1922.
Yaquina Bay (Newport) Bridge, Lincoln County, Three Central Spans, two of which are deck arches, 1936.

STEEL HALF-THROUGH TIED ARCH

Willamette River (Fremont) Bridge, Multnomah County, Single Span, 1973.

STEEL THROUGH TIED ARCH

Clackamas River (McLoughlin) Bridge, Clackmas County, Three Spans, 1933.
South Umpqua River (Winston) Bridge, Douglas County, Three Spans, 1934.
Santiam River bridges, Marion-Linn counties, Three Spans, 1946 and 1958.

Suspension

Willamette River (St. John's) Bridge, Multnomah County, 1931.
Crooked River (Lake Billy Chinook) Bridge, Jefferson County, 1962.

Moveable

STEEL THROUGH TRUSS SWING

Willamette River (Van Buren Street) Bridge, Benton-Linn counties, 1913.
Coquille River Bridge, Coos County, 1922.
Umpqua River (Reedsport) Bridge, Douglas County, 1936.

STEEL THROUGH TRUSS VERTICAL LIFT

Willamette River (Hawthorne) Bridge, Multnomah County, 1910.
Willamette River (Steel) Bridge, Multnomah County, 1912.
Columbia River (Interstate Northbound) Bridge, Multnomah County (Oregon) and Clark County (Washington), 1917.
Columbia River (White Salmon) Bridge, Hood River County (Oregon) and Klickitat County (Washington), 1924.

STEEL SINGLE-LEAF BASCULE

Lewis and Clark River Bridge, Clatsop County, 1924.

STEEL DOUBLE-LEAF BASCULE

Willamette River (Broadway) Bridge, Multnomah County, Through Truss, 1913.
Old Young's Bay Bridge, Clatsop County, 1921.
Willamette River (Burnside) Bridge, Multnomah County, 1926.
Siuslaw River (Florence) Bridge, Lane County, 1936.

Slab, Beam, and Girder

TIMBER DECK GIRDER

Swalley Canal (Rock O' the Range) Covered Bridge, Deschutes County, 1963.

REINFORCED CONCRETE SLAB

Crown Point Viaduct, Multnomah County, 1914.
East Multnomah Falls Viaduct, Multnomah County, 1914.
Horsetail Falls Bridge, Multnomah County, 1914.
Old Mill Race Bridge, Umatilla County, 1914.
Oneonta Gorge Creek (Old) Bridge, Multnomah County, 1914.
Wahkeena Falls Bridge, Multnomah County, 1914.
West Multnomah Falls Viaduct, Multnomah County, 1914.
Johnson Creek Bridge, Multnomah County, 1915.
Ruckel Creek Bridge, Hood River County, 1917.
Gorton Creek Bridge, Hood River County, 1918.
Rock Creek Bridge, Wasco County, 1918.
Chenoweth Creek Bridge, Wasco County, 1920.
Hog Creek Canyon (Rowena Dell) Bridge, Wasco County, 1920.
Beltline Overcrossing, Clatsop County, 1921.

REINFORCED CONCRETE DECK GIRDER

Beaver Creek (Sandy River Overflow) Bridge, Multnomah County, 1912.
Dollarhide Overcrossing, Jackson County, 1914.
Steinman Overcrossing, Jackson County, 1914.
McCarthy Creek Bridge, Multnomah County, Ca. 1914.
Tanner Creek Bridge, Multnomah County, 1915.
Toothrock and Eagle Creek viaducts, Multnomah County, 1915.
Mosier Creek (State Road) Bridge, Wasco County, 1917.
Ruthton Point Viaduct, Hood River County, 1918.
Fifteenmile Creek (Seufert) Viaduct, Wasco County, 1920.
Mill Creek (West Sixth Street) Bridge, Wasco County, 1920.
Rock Slide Viaduct, Hood River County, 1920.
Euchre Creek Bridge, Curry County, 1927.
Pringle Creek (Liberty Street, S.E.) Bridge, Marion County, 1928.
Pringle Creek (Commercial Street, South) Bridge, Marion County, 1928.
Mill Creek (Summer Street, N.E.) Bridge, Marion County, 1929.
Pringle Creek/Shelton Creek (Church Street, S.E.) Bridge, Marion County, 1929.
Elk Creek (Second Crossing) Bridge, Douglas County, 1931.
Link River Bridge, Klamath County, 1931.

Chasm (Neahkahnie Mountain) Bridge, Tillamook County, 1937.
Oneonta Gorge Creek Bridge (New), Multnomah County, 1948.

REINFORCED CONCRETE THROUGH GIRDER

Bridal Veil Falls Bridge, Multnomah County, 1914.
Rhea Creek Bridge, Morrow County, 1916.

STEEL DECK GIRDER

N.E. Grande Avenue Overcrossing, Multnomah County, 1907.
N.E. 12th Avenue Overcrossing, Multnomah County, 1910.
Columbia Slough (N.E. Union Avenue) Bridge, Multnomah County, 1916.
Portland Road, N.E., Undercrossing, Marion County, 1936.
Necarney Creek Bridge, Tillamook County, 1937.

STEEL THROUGH GIRDER

South Umpqua River (Milo Academy) Covered Bridge, Douglas County, 1962.

NUMBER OF BRIDGES BY STRUCTURAL TYPE

The data in this appendix indicate the relative rarity of general and specific highway bridge types in Oregon. The bridge type was assigned on the basis of the main or center span. The historic bridges are those determined eligible for or listed on the National Register and are shown in the body of this report. The nonhistoric bridges are considered ineligible for the National Register. (The reserve bridges are shown in Appendix E). All of the bridges are of pre-1941 construction, except for nine structures--eight covered bridges and the Oneonta Gorge Creek (New) Bridge on the Columbia River Highway.

Type	Historic Bridges	Nonhistoric Bridges Reserve	Nonhistoric Bridges Other
TRUSS	72	24	121
Timber[a]	47	—	—
Deck	1	—	—
Howe	1	—	—
Through (Covered/Housed)	46	—	—
Howe	42	—	—[b]
Kingpost	1	—	—
Queenpost	3	—	—
Iron[a]	2	—	—
Through Pennsylvania-Petit	2	—	—
Steel[c]	23	24	121
Deck	5	7	18
Cantilever	1	—	1
Pratt (Single Span)	1	—	—
Warren	2	3	9
Single Span	1	2	8
Multiple Spans	1	1	1
Warren with Verticals	1	4	8
Single Span	1	4	5
Multiple Spans	—	—	3
Pony	5	9	66
Pratt (Multiple Spans)	—	1	—
Pratt, Half-Hip (Single Span)	2	—	—
Queenpost (Single Span)	1	3	—
Warren (Single Span)	1	2	12
Warren, Double Intersection (Single Span)	—	1	—
Warren with Polygonal Upper Chords	1	2	27
Single Span	—	2	24
Multiple Spans	1	—	3
Warren with Verticals (Single Span)	—	—	5
Warren with Verticals and Polygonal Upper Chords	—	—	22
Single Span	—	—	22
Multiple Spans	—	—	3
Through	13	8	37
Camelback (Single Span)	—	—	3
Cantilever	3	—	—
Parker	3	2	21
Single Span	1	1	17

Type	Historic Bridges	Nonhistoric Bridges Reserve	Other
Multiple Spans	2	1	4
Pennsylvania-Petit	1	1	6
Single Span	—	1	6
Multiple Spans	1	—	—
Pratt	4	3	5
Single Span	2	1	4
Multiple Spans	2	2	1
Warren, Double-Intersection	1	1	1
Single Span	1	—	1
Multiple Spans	—	1	—
Warren with Polygonal Upper Chords (Multiple Spans)	1	1	1[d]
Pin-Connected Metal Truss	11	8	6
Deck	1	—	1
Pony	3	3	—
Through	7	5	5
ARCH	35	11	7
Timber Deck (Single Span)	1	—	—
Reinforced Concrete	30	10	—
Deck	26	8	6
Single Span	19	7	6
Multiple Span	7	1	—
Half-Through (Multiple Spans)	1	—	—
Through (Multiple Spans)	1	—	—
Through Tied	2	2	—
Single Span	1	2	—
Multiple Spans	1	—	—
Steel	4	1	1
Deck (Single Span)	1	—	—
Half-Through (Single Span)	2	—	—
Through (Multiple Spans)	—	—	1
Through Tied	1	1	1
Single Span	—	—	1
Multiple Span	1	1	—
SUSPENSION	1	—	—
MOVEABLE	8	4	6
Bascule	3	2	1
Single-Leaf	—	1	—
Double-Leaf	3	1	1
Removeable/Roll-Back	—	—	2
Swing	2	1	3
Vertical Lift	3	1	—
SLAB, BEAM AND GIRDER[e]	29	14	850[f]
Beam and Girder	17	12	780[f]
Timber	1	—	250[f]
Steel	2	4	180[f]
Reinforced Concrete	14	8	350[f]
Slab, Reinforced Concrete	12	2	70[f]
SUMMARY			
TRUSS	72	24	121
ARCH	35	11	7
SUSPENSION	1	—	—
MOVEABLE	8	4	6
SLAB, BEAM AND GIRDER	29	14	850[f]
TOTAL	145	53	984[f]

(1,182 bridges)

NOTE: This figure provides the number and type of pre-1941 constructed bridges. (The exception is nine structures listed on the National Register that are post-1940 construction, including one on the Columbia River Highway and eight covered bridges.) Since 1940, about 5,250 highway bridges have been built in Oregon, including 135 trusses, 11 arches, 11 moveables, one suspension, and about 5,100 slabs, beams and girders.

^a *All of the timber and iron truss bridges are single spans.*
^b *There are four other Howe timber truss covered bridges in Oregon. The Coast Fork Willamette River (Chambers) Covered Bridge, built in 1936, is a railroad bridge and is therefore excluded from the parameters of this study. Three Howe truss highway covered bridges were constructed after 1940 and were not listed on the National Register in 1979 at the request of the owners. These bridges are Little River (Cavitt Creek), Douglas County, 1943; Crabtree Creek (Bohemian Hall), Linn County, 1947; and Thomas Creek (Skimanek), Linn County, 1966. These bridges are included in the notable post-1940 bridges group in this document (Appendix F).*
^c *This number of steel trusses excludes the trusses found in moveable spans.*
^d *This structure is a Warren with verticals and polygonal upper chords, multiple span.*
^e *Includes both single and multiple spans.*
^f *Estimate.*

APPENDIX E

RESERVE BRIDGES

In the evaluation of Oregon's highway bridges, fifty-three structures were delegated to a reserve category. These structures are not eligible for the National Register, but do exhibit some historic and technological importance. These structures were not considered National Register eligible due primarily to incompatible structural modifications, lack of available historic information, or the existence of better examples of the structure type in the state. In future years, these bridges will be reevaluated against the National Register eligibility criteria. As historic bridges are lost and new historic information is obtained on the reserve structures, some of the reserve bridges may be found National Register eligible.

The fifty-three reserve bridges are listed below, followed by photo-descriptions of the bridges. They are arranged by type, in order: truss, arch, moveable, and slab, beam, and girder. Within each type, the bridges are in chronological order by construction date.

THE RESERVE BRIDGES

Bridge Name	Bridge Number	County	Type	Date
TRUSS:				
Grande Ronde River (Yarrington)	61C16	Union	Steel Through Truss*	1906 (Site-1925)
Greasewood Creek	59C590	Umatilla	Steel Pony Truss (Warren)	1907 (Site-1929)
Rogue River (Robertson)	1592	Josephine	Steel Through Truss* (Parker)	Ca. 1909 (Site-1929)
North Lombard Street Overcrossing	6.2	Multnomah	Steel Deck Truss (Warren)	1909
Jordan Creek Overflow	45C601	Malheur	Steel Pony Truss* (Queenpost)	Ca. 1910 (Site-1932)
Pine Creek	59C528	Umatilla	Steel Pony Truss* (Queenpost)	Ca. 1910 (Site-1924)
Pine Creek	10712	Umatilla	Steel Pony Truss* (Queenpost)	Ca. 1910 (Site-1936)
South Umpqua River (Worthington)	7408	Douglas	Steel Through Truss* (Pennsylvania-Petit)	Ca. 1910 (Site-1951)
South Yamhill River (Steel)	53C05	Polk	Steel Through Truss* (Pratt)	Ca. 1910
Pine Creek (Crow Lane)	1C826	Baker	Steel Pony Truss (Warren)	1911
Crooked River (Elliott Lane)	7C12	Crook	Steel Through Truss (Double-Intersection Warren)	Ca. 1914 (Site-1936)
Marys River	14478	Benton	Steel Pony Truss (Warren with Polygonal Upper Chords)	1914 (Site-1938)
Beaver Creek (Kukkla Road)	9C83	Columbia	Steel Pony Truss (Double-Intersection Warren)	Ca. 1915 (Site-1964)
Luckiamute River (Grant Road)	53C139	Polk	Steel Through Truss* (Pratt)	1918 (Site-1958)
Powder River (Myrtle Street)	4T04	Baker	Steel Pony Truss (Pratt)	1920

*Pin-connected truss.

Clackamas River (Park Place)	604	Clackamas	Steel Through Truss (Parker)	1921
North Yamhill River	441	Yamhill	Steel Deck Truss (Warren with Verticals)	1921
Calapooya Creek (Oakland)	603	Douglas	Steel Deck Truss (Warren)	1925
Grande Ronde River (McKennon Road)	61C19	Union	Steel Pony Truss (Warren with Polygonal Upper Chords)	1925
Deschutes River (Maupin)	966	Wasco	Steel Deck Truss (Warren)	1929
Umpqua River (Scottsburg)	1318	Douglas	Steel Through Truss (Warren with Polygonal Upper Chords)	1929
Elk Creek (First Crossing)	1614	Douglas	Steel Deck Truss (Warren with Verticals)	1931
Elk Creek (Third Crossing)	1465	Douglas	Steel Deck Truss (Warren with Verticals)	1931
Elk Creek (Fourth Crossing)	1406	Douglas	Steel Deck Truss (Warren with Verticals)	1931

ARCH:

Grande Ronde River (Perry Overcrossing)	626	Union	Reinforced Concrete Deck Arch	1924
Umatilla River (Umatilla)	624A	Umatilla	Reinforced Concrete Deck Arch	1925
Soapstone Creek	1319	Clatsop	Reinforced Concrete Deck Arch	1928
Salmon River	4192	Lincoln	Reinforced Concrete Deck Arch	1930
Big Creek	1180	Lane	Reinforced Concrete Through Tied Arch	1931
Cummins Creek	1182	Lane	Reinforced Concrete Deck Arch	1931
Tenmile Creek	1181	Lane	Reinforced Concrete Through Tied Arch	1931
Hood River (Tucker)	1600	Hood River	Reinforced Concrete Deck Arch	1932
South Umpqua River (Winston)	1923	Douglas	Steel Through Tied Arch	1934
Clackamas River (Estacada)	2208	Clackamas	Reinforced Concrete Deck Arch	1936
Nehalem River	2165	Clatsop	Reinforced Concrete Deck Arch	1939

MOVEABLE:

Willamette River (Van Buren Street)	2728	Benton-Linn	Steel Through Truss (Pratt) Swing	1913
Old Young's Bay	330	Clatsop	Steel Double-Leaf Bascule	1921
Columbia River (White Salmon)	6645	Hood River	Steel Through Truss (Pennsylvania-Petit) Vertical Lift	1924
Lewis and Clark River	711	Clatsop	Steel Single-Leaf Bascule	1924

SLAB, BEAM, AND GIRDER:

N.E. Grand Avenue Overcrossing	7040	Multnomah	Steel Deck Girder	1907
N.E. 12th Avenue Overcrossing	7039	Multnomah	Steel Deck Girder	1910
McCarthy Creek	—	Multnomah	Reinforced Concrete Deck Girder	Ca. 1914
Johnson Creek	51C02	Multnomah	Reinforced Concrete Slab	1915
Columbia Slough (N.E. Union Avenue)	1377C	Multnomah	Steel Deck Girder	1916
Rhea Creek	49C23	Morrow	Reinforced Concrete Through Girder	1916
Mosier Creek (State Road)	118	Wasco	Reinforced Concrete Deck Girder	1917

Beltline Overcrossing	*2418*	*Clatsop*	*Reinforced Concrete Slab*	*1921*
Euchre Creek Bridge	*15C31*	*Curry*	*Reinforced Concrete Deck Girder*	*1927*
Pringle Creek (Commercial Street South)	*1340*	*Marion*	*Reinforced Concrete Deck Girder*	*1928*
Mill Creek (Summer Street, N.E.)	*1357S*	*Marion*	*Reinforced Concrete Deck Girder*	*1929*
Elk Creek (Second Crossing)	*1601*	*Douglas*	*Reinforced Concrete Deck Girder*	*1931*
Link River	*1579*	*Klamath*	*Reinforced Concrete Deck Girder*	*1931*
Portland Road, N.E., Undercrossing	*2131*	*Marion*	*Steel Deck Girder*	*1936*

TRUSS

Grande Ronde River (Yarrington) Bridge
Structure Number 61C16
Constructed - 1906 (Site-1925)
Steel Through Truss (Pratt)
Yarrington-Scott County Road 49
Palmer Junction, Union County
Ownership - Union County

Moved to this site in 1925, this two-span steel Pratt truss is an example of the pin-connected construction technology used prior to about 1915. The structure is 250 feet long, consisting of 150-foot and 100-foot through truss spans. A name plaque attached to the 100-foot span indicates that it was built in 1906 by P.S. Easterday and Company of Walla Walla.

Greasewood Creek Bridge
Structure Number 59C590
Constructed - 1907 (Site-1929)
Steel Pony Truss (Warren)
County Road 854
Adams vicinity, Umatilla County
Ownership - Umatilla County

This 60-foot steel Warren pony truss is located on a lightly-traveled county road in the rolling farmland of northeast Oregon. It is the oldest Warren truss on Oregon's highway system and also is the earliest example in the state with riveted truss connections. The structure has a timber roadway deck and is constructed from relatively light-gauge angle steel. The sway braces are an unusual feature on this pony truss. Half of the original name plaque is missing, leaving only the first names of the local officials from the period. The bridge was built by P.S. Easterday Company of Walla Walla, Washington.

Rogue River (Robertson) Bridge
Structure Number 1592
Constructed - Ca. 1909 (Site-1929)
Steel Through Truss (Parker)
Rogue River Loop Highway 260, M.P. 12.91
Merlin Vicinity, Josephine County
Ownership - State of Oregon

The Robertson Bridge, named for early Oregon pioneers who settled near the area in the 1870s, is a 583-foot three-span steel through truss structure with one 90-foot Pratt truss, two 180-foot Parker trusses, and seven timber frame approach spans. The three truss spans are all pin-connected. Twenty-five pin-connected truss highway bridges remain in Oregon. The bridge was moved to its present site in 1929 from a location on the Pacific Highway over the Rogue River in Grants Pass, where it was replaced by the Caveman Bridge (1931). The county officials listed on the nameplate provide a dating of the structure, circa 1909.

North Lombard Street Overcrossing
Structure Number 6.2
Constructed - 1909
Steel Deck Truss (Warren)
North Lombard Street
Portland, Multnomah County
Ownership - Burlington Northern Railroad

Constructed in 1909 by the Spokane, Portland, and Seattle Railroad, this structure consists of three 90-foot hanging Warren truss spans supported on steel towers. The truss members are rivet connected. Similar to two neighboring truss bridges on North Fessenden Street and North Willamette Boulevard, the structure was designed by noted engineer Ralph Modjeski. The bridge has a vintage lattice steel handrail. (Additional information about the Spokane, Portland, and Seattle Railroad project in North Portland is in the description of the North Fessenden Street Overcrossing.)

Jordan Creek Overflow Bridge
Structure Number 45C601
Constructed - Ca. 1910 (Site-1932)
Steel Pony Truss (Queenpost)
County Road 808
Jordan Valley vicinity, Malheur County
Ownership - Malheur County

This structure is a 32-foot pony truss span of queenpost configuration. The bridge is pin connected, indicating a construction date of about 1910. Records show the bridge was moved to the site in 1932. The bridge has a timber roadway deck and is located on a lightly-traveled county road. This truss bridge is one of very few in Oregon supported by masonry abutments.

Pine Creek Bridge
Structure Number 59C528
Constructed - Ca. 1910 (Site-1924)
Steel Pony Truss (Queenpost)
County Road 707
Umapine vicinity, Umatilla County
Ownership - Umatilla County

This structure is a 40-foot steel queenpost pony truss. The bridge is one of four queenpost trusses on Oregon's highway system, one of six pin-connected pony trusses and one of twenty-five highway trusses (pony and through) using pin-connections. The bridge was installed at its current site in 1924.

Pine Creek Bridge
Structure Number 10712
Constructed - Ca. 1910 (Site-1936)
Steel Pony Truss (Queenpost)
County Road 676
Weston vicinity, Umatilla County
Ownership - Umatilla County

The Pine Creek Bridge is a 40-foot steel pony truss of queenpost configuration. Its construction date is established on the basis of the pin-connected technology, largely obsolete by 1915. It is one of only four queenpost trusses on Oregons highway system and one of twenty-five pin-connected trusses. The bridge was moved to this site in 1936.

South Umpqua River (Worthington) Bridge
Structure Number 7408
Constructed - Ca. 1910 (Site-1951)
Steel Through Truss (Pennsylvania-Petit)
County Road 42, M.P. 8.82
Days Creek, Douglas County
Ownership - Douglas County

Known locally as the Worthington Bridge, this bridge is one of Oregon's twenty-five pin-connected highway truss spans. The bridge was moved from its original location on the Rogue River to this site in 1951, when it was modified from 190 feet to its current length of 153 feet. Though its construction date is unknown, the use of pin-connection technology indicates a construction date of about 1910.

South Yamhill River (Steel) Bridge
Structure Number 53C05
Constructed - Ca. 1910
Steel Through Truss (Pratt)
County Road 669 (Closed)
Willamina, Polk County
Ownership - Polk County

Two 149-foot steel through truss spans make up this 325-foot long bridge. The Pratt trusses are connected with pins. The one lane bridge is closed to vehicular traffic.

Pine Creek (Crow Lane) Bridge
Structure Number 1C826
Constructed - 1911
Steel Pony Truss (Warren)
Crow Lane County Road 1027
Halfway vicinity, Baker County
Ownership - Baker County

This 53-foot steel structure is one of the oldest Warren pony trusses on Oregon's highway system. It is also one of the oldest bridges in Oregon with truss members connected with rivets.

Crooked River (Elliott Lane) Bridge
Structure Number 7C12
Constructed - Ca. 1914 (Site-1936)
Steel Through Truss (Double-Intersection Warren)
Elliott Lane County Road 124
Prineville vicinity, Crook County
Ownership - Crook County

One of only two steel through trusses of double-intersection Warren configuration remaining on Oregon's highway system, the Elliott Lane Bridge consists of a 125-foot through truss main span and an 80-foot Warren with polygonal upper chords pony truss. The truss members are connected with rivets. The through truss was probably constructed by the Coast Bridge Company in 1914, relocated, and lengthened (with the pony truss addition) in 1936.

Marys River Bridge
Structure Number 14478
Constructed - 1914 (Site-1938)
Steel Pony Truss (Warren with Polygonal
Upper Chords)
Harris County Road 16520
Alder vicinity, Benton County
Ownership - Benton County

Built by the Coast Bridge Company of Portland in 1914, this 138-foot steel pony truss was moved to its present site in 1938. The truss is rivet-connected and a Warren with polygonal upper chords.

Beaver Creek (Kukkla Road) Bridge
Structure Number 9C83
Constructed - Ca. 1915 (Site-1964)
Steel Pony Truss
(Double-Intersection Warren)
Kukkla County Road 4087
Clatskanie vicinity, Columbia County
Ownership - Columbia County

The Kukkla Road Bridge is a 48-foot, rivet-connected steel pony truss. It is the only pony truss in Oregon with a double-intersection Warren truss. The structure was purchased by the county from the state and moved to this location in 1964. Its original location is believed to have been near the Oregon Coast.

* Luckiamute River (Grant Road) Bridge
Structure Number 53C139
Constructed - 1918 (Site-1958)
Steel Through Truss (Pratt)
Grant County Road 1062
* Kings Valley vicinity, Polk County
Ownership - Polk County

This 140-foot pin-connected Pratt truss span is one of twenty-five remaining pin-connected highway trusses in Oregon. Originally spanning the Luckiamute River at another location in the northern part of Polk County, this 1918 bridge was relocated to its present site in 1958. The steel grate decking is from the old Morrison Street Bridge (1905), Portland.

* This bridge was removed in 1987.

Powder River (Myrtle Street) Bridge *
Structure Number 4T04
Constructed - 1920
Steel Pony Truss (Pratt)
Myrtle Street *
Baker, Baker County
Ownership - Baker County

This 65-foot steel pony truss is a Pratt with rivet connections. It was built by the City of Baker and displays diagonal bracing of the vertical truss members to increase lateral strength in the event of high water.

* *The Powder River Bridge was removed in 1987.*

Clackamas River (Park Place) Bridge
Structure Number 604
Constructed - 1921
Steel Through Truss (Parker)
82nd Avenue (Closed)
Gladstone, Clackamas County
Ownership - City of Gladstone

Originally a major crossing of the Clackamas River on the route from Oregon City to Portland, this 220-foot steel through truss span has been closed to traffic since 1976, but is being reopened for bicycle and pedestrian traffic. The truss is a Parker with rivet connections.

North Yamhill River Bridge
Structure Number 441
Constructed - 1921
Steel Deck Truss (Warren with Verticals)
Pacific Highway West 1W, M.P. 34.99
McMinnville vicinity, Yamhill County
Ownership - State of Oregon

An 80-foot steel deck truss (Warren with verticals) and seven 40-foot concrete deck girder spans make up this 360-foot bridge. The truss is rivet connected. The structure was designed under the auspices of Conde B. McCullough, State Bridge Engineer.

Calapooya Creek (Oakland) Bridge
Structure Number 603
Constructed - 1925
Steel Deck Truss (Warren)
Oakland-Shady Highway 324, M.P. 1.08
Oakland, Douglas County
Ownership - Douglas County

Located at Oakland on old Highway 99, this structure consists of a 100-foot steel Warren deck truss and nine concrete deck girder approach spans. The total length of the bridge is 473 feet. The precast concrete railing, bracketing, and overall configuration are representative of a standard state bridge design of the 1920s.

Grande Ronde River Bridge
(McKennon Road)
Structure Number 61C19
Constructed - 1925
Steel Pony Truss (Warren with Polygonal Upper Chords)
McKennon County Road 132
Imbler vicinity, Union County
Ownership - Union County

Lying in the farmland southeast of Imbler, this bridge is a 100-foot steel pony truss. The truss is a Warren with polygonal upper chords and is rivet-connected. The roadway deck is timber, and the support piers are steel caissons.

Deschutes River (Maupin) Bridge
Structure Number 966
Constructed - 1929
Steel Deck Truss (Warren)
The Dalles-California Highway 4, M.P. 45.87
Maupin vicinity, Wasco County
Ownership - State of Oregon

The Deschutes River Bridge is an 826-foot structure, consisting of a 200-foot steel Warren deck truss and thirteen concrete girder approach spans. The concrete towers, the ornate bridge railing, and soffit bracketing add aesthetic interest to the structure. The structure was designed by Conde B. McCullough and built by the Kuckenberg-Wittman Company.

Umpqua River (Scottsburg) Bridge
Structure Number 1318
Constructed - 1929
Steel Through Truss (Warren with Polygonal Upper Chords)
Umpqua Highway 45, M.P. 16.43
Scottsburg, Douglas County
Ownership - State of Oregon

Three spans form this 643-foot continuous through truss bridge at the historic community of Scottsburg. Designed by Conde B. McCullough, it is very similar to the Willamette River Bridge at Springfield, also constructed in 1929. The Scottsburg Bridge was built by the Clackamas Construction Company.

Elk Creek (First Crossing) Bridge
Structure Number 1614
Constructed - 1931
Steel Deck Truss (Warren with Verticals)
Umpqua Highway 45, M.P. 36.39
Elkton vicinity, Douglas County
Ownership - State of Oregon

This 140-foot steel deck truss contains a Warren with verticals truss. This bridge, along with three other associated Elk Creek crossings, was designed by Conde B. McCullough and was constructed on the Umpqua Highway in 1931. It is the only one of the four associated Elk Creek crossings to carry a commemorative dedication. A bronze plaque dedicates this structure to Anna Elizabeth Wells.

Elk Creek (Third Crossing) Bridge
Structure Number 1465
Constructed - 1931
Steel Deck Truss (Warren with Verticals)
Umpqua Highway 45, M.P. 39.64
Elkton vicinity, Douglas County
Ownership - State of Oregon

One of the four Elk Creek crossings built in 1931 on the Umpqua Highway, this bridge, the adjacent Elk Creek Tunnel, and the Elk Creek (4th Crossing) Bridge form almost a continuous structure. This structure, designed by Conde B. McCullough, is a 140-foot steel deck truss (Warren with verticals). It displays an ornate bridge railing, as do the other three Elk Creek crossings.

Elk Creek (Fourth Crossing) Bridge
Structure Number 1406
Constructed - 1931
Steel Deck Truss (Warren with Verticals)
Umpqua Highway 45, M.P. 39.97
Elkton vicinity, Douglas County
Ownership - State of Oregon

This 100-foot steel deck truss is a Warren with verticals. Its design is attributed to Conde B. McCullough, State Bridge Engineer. The bridge is one of four Elk Creek crossings constructed in 1931 on the Umpqua Highway. (The second crossing structure is a reinforced concrete deck girder and is shown later in this appendix.) Portions of the original ornamental sidewalk railing have been replaced.

ARCH

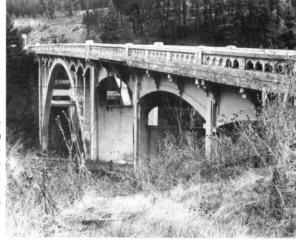

Grande Ronde River Bridge
(Perry Overcrossing)
Structure Number 626
Constructed - 1924
Reinforced Concrete Deck Arch
Old Oregon Trail Highway 6, M.P. 256.31
(Frontage Road)
La Grande vicinity, Union County
Ownership - State of Oregon

The Perry Overcrossing, another of Conde B. McCullough's arch bridges, was built on the Hilgard-La Grande section of the Old Oregon Trail Highway. The Union Bridge Company was awarded the contract for this 312-foot long bridge. The main span is a 134-foot reinforced concrete rib deck arch over the Grande Ronde River. The open spandrel arch has curved arch fascia curtain walls, and the piers have partially solid web walls. Small arched openings accentuate the railing, and the railing soffits are ornately curved. The still-intact blast plates above the railroad tracks originally protected the underside of the structure from the intense heat and gases of passing steam engines.

Umatilla River (Umatilla) Bridge
Structure Number 624A
Constructed - 1925
Reinforced Concrete Deck Arch
Columbia River Highway 2, M.P. 182.60
Umatilla, Umatilla County
Ownership - State of Oregon

Dedicated to an early member of the State Highway Commission, William Duby, this 439-foot structure consists of three 110-foot open-spandrel, rib-type, reinforced concrete deck arch spans, and six concrete deck girder approach spans. Designed by Conde B. McCullough, the structure has arched spandrel curtain walls with bush-hammered inset panels. The original decorative concrete railing was removed during a widening project in 1951 and was replaced with a galvanized steel railing typical of the 1950s period.

Soapstone Creek Bridge
Structure Number 1319
Constructed - 1928
Reinforced Concrete Deck Arch
Necanicum Highway 46, M.P. 6.50
Necanicum Junction vicinity,
Clatsop County
Ownership - State of Oregon

One of Conde B. McCullough's arch bridges, this structure replaced a wooden covered bridge. The contract for the bridge was awarded to Lindstrom and Feigenson of Portland. The 152-foot structure exhibits many features characteristic of McCullough bridges, including an open spandrel with curved arched fascia curtain walls, precast railings with small arched openings and topped with concrete caps, and curved soffits. When the Soapstone Bridge was constructed in 1928, the present Necanicum Highway was the route of the Oregon Coast Highway.

Salmon River Bridge
Structure Number 4192
Constructed - 1930
Reinforced Concrete Deck Arch
Salmon River Highway 39, M.P. 6.23
Rose Lodge vicinity, Lincoln County
Ownership - State of Oregon

Constructed by the United State Bureau of Public Roads, the main piers and rib members of this bridge are skewed to allow for passage of the Salmon River. The main span is an 80-foot open-spandrel, rib-type reinforced concrete deck arch. The railing is ornamental, and the spandrel columns curve at the top to form decorative brackets to support the sidewalk and railing. The bridge was designed by H.R. Angwuir, Senior Bridge Engineer, United State Bureau of Public Roads, San Francisco.

Big Creek Bridge
Structure Number 1180
Constructed - 1931
Reinforced Concrete Through Tied Arch
Oregon Coast Highway 9, M.P. 175.02
Heceta Head vicinity, Lane County
Ownership - State of Oregon

The main span of the Big Creek Bridge is a 120-foot reinforced concrete through tied arch, elliptical in shape. Very similar in design and size to the Wilson River and Tenmile Creek bridges, the Big Creek Bridge is one of the first reinforced concrete tied arch spans to be constructed in this country. The bridge is well suited to its location and has an attractive precast concrete railing. The concrete deck girder approach spans have arched fascia curtain walls and contribute to an overall structure length of 235 feet. The bridge was designed by Conde B. McCullough, State Bridge Engineer, and constructed by the Union Bridge Company.

Cummins Creek Bridge
Structure Number 1182
Constructed - 1931
Reinforced Concrete Deck Arch
Oregon Coast Highway 9, M.P. 178.35
Yachats vicinity, Lane County
Ownership - State of Oregon

Located about four miles south of Yachats on the Oregon Coast Highway, the bridge spans Cummins Creek at Neptune State Park. The bridge's total length is 185 feet, and the main span is a 115-foot reinforced concrete rib arch. The arch is an open spandrel type with a low rise. The arch appears to be segmented because of block recessed lines. The main piers and spandrel columns are fluted, as are the railing posts. The railing is supported by curved brackets and consists of small semi-circular arched openings. The structure was designed by Conde B. McCullough, State Bridge Engineer, and built by Tom Lillebo, Contractor.

Tenmile Creek Bridge
Structure Number 1181
Constructed - 1931
Reinforced Concrete Through Tied Arch
Oregon Coast Highway 9, M.P. 171.44
Yachats vicinity, Lane County
Ownership - State of Oregon

Located in an attractive stretch of Oregon's coastline, the Tenmile Creek Bridge was designed by Conde B. McCullough, State Bridge Engineer. Along with the Wilson River and Big Creek bridges, the Tenmile Creek Bridge illustrates the first uses of the reinforced concrete tied arch in the United States. The main span is 120 feet, and the total length is 180 feet. The structure has ornate precast concrete railings.

Hood River (Tucker) Bridge
Structure Number 1600
Constructed - 1932
Reinforced Concrete Deck Arch
Hood River Highway 281, M.P. 4.95
Hood River vicinity, Hood River County
Ownership - State of Oregon

This structure spans Hood River about four miles south of the city of Hood River. The bridge carries the historic name of Tucker's Bridge. B.R. Tucker built a bridge and sawmill at this located about 1881. The 1932 arch bridge was designed by the State Bridge Engineer Conde B. McCullough, but was built by Hood River County. The bridge was contracted to Charles O. Young. Totaling 188 feet in length, the main span is a 100-foot reinforced concrete rib deck arch. The spandrel columns, main piers and approach columns are slender and have recessed narrow vertical panes. The spandrel columns and approach piers also have a shallow arched fascia curtain walls. The railings are supported by curved brackets and display a series of semi-circular arched openings.

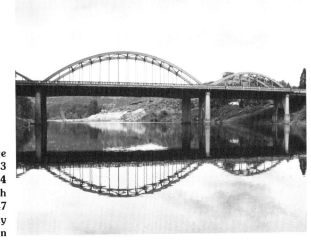

South Umpqua River (Winston) Bridge
Structure Number 1923
Constructed - 1934
Steel Through Tied Arch
Coos Bay-Roseburg Highway 35, M.P. 74.47
Winston vicinity, Douglas County
Ownership - State of Oregon

The Winston Bridge consists of three steel tied arch spans, including a central arch of 180 feet, flanked by two smaller 100-foot arches. There are four concrete deck girder approach spans, giving a total length of 548 feet. The structure was designed by Conde B. McCullough and has an ornate precast concrete railing and arch openings between the bridge piers.

Clackamas River (Estacada) Bridge
Structure Number 2208
Constructed - 1936
Reinforced Concrete Deck Arch
Woodburn-Estacada Highway 161,
M.P. 33.40
Estacada, Clackamas County
Ownership - State of Oregon

The main span of this 371-foot structure is a 140-foot open-spandrel, rib-type reinforced concrete deck arch between eight concrete deck girder approach spans. The main arch has a very high rise. The roadway is skewed on the structure. The structure was designed by Conde B. McCullough and is embellished with a decorative precast concrete railing, elaborate projecting brackets, and lean, tall spandrel columns. This bridge replaced a timber Howe truss covered bridge at this site. The arch bridge was constructed by the Mountain States Construction Company, Eugene.

Nehalem River Bridge
Structure Number 2165
Constructed - 1939
Reinforced Concrete Deck Arch
Sunset Highway 47, M.P. 21.73
Elsie vicinity, Clatsop County
Ownership - State of Oregon

    This 617-foot long structure consists of a 231-foot open-spandrel, rib-type reinforced concrete deck arch, and eleven concrete deck girder approach spans. The approach and spandrel columns have semi-circular arched fascia curtain walls. The approach piers also have arched fascia walls. The slimness of the arch ribs, piers, and columns contribute to the attractive appearance of the bridge. Long brackets support the ornamental bridge railings and sidewalks. The designer of the bridge was Glenn S. Paxson, State Bridge Engineer. The bridge was built by the Mountain States Construction Company, Eugene.

MOVEABLE

Willamette River (Van Buren Street) Bridge
Structure Number 2728
Constructed - 1913
Steel Through Truss (Pratt) Swing
Corvallis-Lebanon Highway 210, M.P. 0.13
Corvallis, Benton-Linn counties
Ownership - State of Oregon

This three-span structure is the oldest swing-span bridge in Oregon and is the only pin-connected swing-span truss. It was constructed by the Coast Bridge Company and consists of a 249-foot steel through truss swing span, a 171-foot steel through truss, a 57-foot steel pony truss, and nine 19-foot timber spans. The mechanism to operate the swing span was removed in the 1950s, and there is little decorative or architectural treatment on the structure. The bridge was originally built by Benton County, but now belongs to the state.

Old Young's Bay Bridge
Structure Number 330
Constructed - 1921
Steel Double-Leaf Bascule
Warrenton Highway 105, M.P. 6.89
Astoria vicinity, Clatsop County
Ownership - State of Oregon

This steel double-leaf bascule drawspan structure was the first moveable span bridge designed by state bridge engineer Conde B. McCullough. The 150-foot central bascule span consists of two 75-foot central cantilevered sections operated by 40 horsepower electric motors and counterweights. Fifty-eight pile trestle secondary spans and ten timber stringer spans carry 1,616 feet of approach roadway to the central span and contribute to an overall structure length of 1,766 feet. Typical of McCullough designs, large ornate approach pylons with lanterns stand on either side of the roadway at both ends of the structure.

Columbia River (White Salmon) Bridge
Structure Number 6645
Constructed - 1924
Steel Through Truss (Pennsylvania-Petit)
Vertical Lift
Port Road connecting with Columbia River
Highway 2, M.P. 64.62
Hood River, Hood River County (Oregon) and
White Salmon, Klickitat County (Washington)
Ownership - Port of Hood River

The White Salmon Bridge is the second oldest highway bridge across the Columbia River between Oregon and Washington. The 4,755-foot long toll bridge consists of a 262-foot steel through Pennsylvania-Petit truss vertical lift span and sixteen 206-foot steel deck truss secondary spans. The bridge, originally called the Waucoma Interstate Bridge, was constructed by Gray and Chandler, under the direction of Professor C.B. Wing of Stanford University, for the Oregon-Washington Bridge Company. The bridge was originally not a moveable span structure, but was converted to a vertical lift type in 1939 in conjunction with the completion of the Bonneville Dam downstream.

Lewis and Clark River Bridge
Structure Number 711
Constructed - 1924
Steel Single-Leaf Bascule
Warrenton Highway 105, M.P. 4.78
Astoria vicinity, Clatsop County
Ownership - State of Oregon

The Lewis and Clark River Bridge is the only remaining single-leaf bascule draw span built before World War II. The central span is a 112-foot steel single-leaf bascule span operated by dual electric motors and provides 105 feet of lateral waterway clearance. Forty-eight pile trestle and stringer spans carry 716 feet of approach roadway, contributing to a total structure length of 828 feet. Designed by Conde B. McCullough, the bridge was constructed, along with the Old Young's Bay Bridge, as part of an improvement project for the coastal area.

SLAB, BEAM, AND GIRDER

N.E. Grand Avenue Overcrossing
Structure Number 7040
Constructed - 1907
Steel Deck Girder
Pacific Highway East 1E, M.P. 0.20
Portland, Multnomah County
Ownership - State of Oregon

This seven-span, 352-foot steel deck girder is the fifth oldest highway bridge in Oregon and the oldest slab, beam, and girder type structure. A plaque mounted on the ornate lattice-steel railing indicates the bridge was built by the International Contract Company of Seattle, Washington. The original decorative lampposts and lanterns are on the bridge. This was the first major span across Sullivan's Gulch. Interstate 84 and the Union Pacific Railroad pass under the structure.

N.E. 12th Avenue Overcrossing
Structure Number 7039
Constructed - 1910
Steel Deck Girder
N.E. 12th Avenue, over Interstate 84
and the Union Pacific Railroad
Portland, Multnomah County
Ownership - City of Portland

Built in 1910 as the third major crossing of Sullivan's Gulch, this structure is a 320-foot steel deck girder on steel towers. The main span of the bridge was altered during the construction of the Banfield Freeway in 1956, but the work was smoothly integrated into the original design. The vintage lattice steel sidewalk railing was retained. The structure was designed by Waddell and Harrington, Kansas City, Missouri, and was constructed by the International Contract Company, Seattle.

McCarthy Creek Bridge
Structure Number Unknown
Constructed - Ca. 1914
Reinforced Concrete Deck Girder
Lower Columbia River Highway 2W,
M.P. 13.19 (Bypassed)
Portland vicinity, Multnomah County
Ownership - State of Oregon

    Little information has been obtained on this abandoned concrete deck girder span northwest of Portland. It was originally constructed on the Lower Columbia River Highway, but now lies abandoned next to a new bridge.

Johnson Creek Bridge
Structure Number 51C02
Constructed - 1915
Reinforced Concrete Slab
S.E. Tacoma Street
Portland, Multnomah County
Ownership - City of Portland

    This structure is a 28-foot concrete slab span built in 1915 and located at the south end of the Eastmoreland Golf Course in southeast Portland. The structure is joined to a large curved retaining wall which is topped by an ornate spindle balustrade railing. It is not known whether this is part of the original structure or added at a later date.

Columbia Slough (N.E. Union Avenue) Bridge
Structure Number 1377C
Constructed - 1916
Steel Deck Girder
N.E. Union Avenue Crossing of the
Pacific Highway East 1E, M.P. 4.41
Portland, Multnomah County
Ownership - State of Oregon

Consisting of four 76-foot steel deck girder spans, this 304-foot structure was constructed in 1916 as part of the approach roadway leading to the Interstate Bridge (1917) over the Columbia River. It is very similar to the Oregon Slough Bridge on Interstate 5 at the Interstate Bridge, which was built at the same time and which exhibits the same ornate vintage lattice steel railing.

Rhea Creek Bridge
Structure Number 49C23
Constructed - 1916
Reinforced Concrete Through Girder
Ruggs-Jordan County Road 581
Jordan vicinity, Morrow County
Ownership - Morrow County

Located on a lightly-traveled unpaved county road, this is one of only two known concrete through girder spans on Oregon's highway system. (The other is the Bridal Veil Falls Bridge, constructed in 1914 on the old Columbia River Highway.) The structure is 42 feet long. The end post provides the name of the builder, John W. Ash, and the date of construction. The design for the bridge was provided to the county by the State Highway Department.

Mosier Creek (State Road) Bridge *
Structure Number 118
Constructed - 1917
Reinforced Concrete Deck Girder
State (County) Road
Mosier vicinity, Wasco County *
Ownership - Wasco County

This 175-foot structure consists of five 35-foot reinforced concrete girder spans. The structure is given aesthetic interest by the spindle-type balustrade railing. This bridge was designed by the state, but built by Wasco County. The bridge plans are signed by State Highway Department bridge designer L.W. Metzger. The bridge was built concurrent with the construction of the old Columbia River Highway in Wasco County, but was not on the route of the Columbia River Highway.

* *The Mosier Creek Bridge was removed in 1987.*

Beltline Overcrossing
Structure Number 2418
Constructed - 1921
Reinforced Concrete Slab
Warrenton Highway 105, M.P. 7.10
Astoria vicinity, Clatsop County
Ownership - State of Oregon

Originally constructed as a railroad overpass, this bridge is the most ornate concrete slab structure in Oregon. It consists of two 18-foot reinforced concrete slab spans which have arched fascia walls with keystones. Arches, columns, and keystones embellish the wing walls. The precast concrete railing is a typical 1920s design. The structure is located near the old Young's Bay Bridge, built in the same year.

Euchre Creek Bridge
Structure Number 15C31
Constructed - 1927
Reinforced Concrete Deck Girder
Ophir County Road 510
Ophir, Curry County
Ownership - Curry County

This structure is a 90-foot reinforced concrete deck girder, consisting of three 30-foot spans. The arched girder members with bush-hammered insets, soffit brackets, and precast arched concrete railing make this a good example of late 1920s bridge construction. This bridge was originally on the Oregon Coast Highway, but is now on a bypassed section.

Pringle Creek Bridge
(Commercial Street, South)
Structure Number 1340
Constructed - 1928
Reinforced Concrete Deck Girder
Pacific Highway East 1E, M.P. 50.59
Salem, Marion County
Ownership - State of Oregon

One of the bridges built during Salem's bridge construction program of 1928-29, this 192-foot structure consists of four 48-foot reinforced concrete deck girder spans and features the same ornamental railing, brackets, and arched girder members which appear on all the bridges built during the program. East of the bridge is the landscaped grounds of Salem's Civic Center and westward is the Boise-Cascade Paper Mill.

Mill Creek (Summer Street, N.E.) Bridge
Structure Number 1357S
Constructed - 1929
Reinforced Concrete Deck Girder
Pacific Highway East 1E, M.P. 49.28
Salem, Marion County
Ownership - State of Oregon

This small bridge exhibits the ornamental bridge railing, arched girder members, bush-hammered insets, and bracketing common to most of the bridges constructed during Salem's bridge construction program of 1928-29. The Summer Street Bridge is a 60-foot, reinforced concrete deck girder, consisting of three 20-foot spans. The bridge is located several blocks north of the State Capitol.

Elk Creek (Second Crossing) Bridge
Structure Number 1601
Constructed - 1931
Reinforced Concrete Deck Girder
Umpqua Highway 45, M.P. 38.76
Elkton vicinity, Douglas County
Ownership - State of Oregon

Unlike the other three Elk Creek crossings built on the Umpqua Highway, the second crossing structure is a concrete deck girder, while the others are steel deck trusses. This structure consists of six spans totaling 290 feet and has the same ornate bridge railing found on the other Elk Creek bridges. Design by Conde B. McCullough, the structure was built by Odom and Frazier, Contractors.

Link River Bridge
Structure Number 1579
Constructed - 1931
Reinforced Concrete Deck Girder
Klamath Falls-Lakeview Highway 20,
M.P. 0.02
Klamath Falls, Klamath County
Ownership - State of Oregon

Fluted entrance pylons, graceful arched girder members, bracketing, and ornate railings are architectural features on this 200-foot concrete deck girder bridge in Klamath Falls. The four-span structure was designed by the State Highway Department and constructed by Lindstrom and Feigenson, Contractors. On the approach railing at the east end of the bridge is a bronze plaque commemorating the founding of Linkville, later called Klamath Falls, founded in 1867 by pioneer George Nurse.

Portland Road, N.E., Undercrossing
Structure Number 2131
Constructed - 1936
Steel Deck Girder
Pacific Highway East 1E, M.P. 47.78
Salem, Marion County
Ownership - State of Oregon

Designed by Conde B. McCullough, this underpass has decorative fluted pylons, two pedestrian tunnels, and attractive gothic-arch balustrade railings typical of Depression-era structures. The main structure is a 49-foot steel deck girder span over 100 feet wide, its large width due to the skewed approach of the railroad tracks to the roadway.

APPENDIX F

NOTABLE HIGHWAY BRIDGES CONSTRUCTED AFTER 1940

The historic bridges study was designed in accordance with the basic eligibility requirements for National Register listing. One of these requirements is that properties be at least fifty years old, unless of exceptional significance. The historic bridges study systematically inventoried Oregon's highway bridges built prior to 1941. Although this cutoff date includes some bridges a few years younger than fifty years, it was established to keep the study results current for several years and to allow for advanced planning. The study did, in addition, examine several notable bridges constructed after 1940 to test for exceptional historic significance.

Eight bridges were identified as notable post-1940 construction bridges. The study review team found none of these structures of exceptional historic significance and concluded that they were presently ineligible for the National Register.

The post-1940 bridges are identified below and are shown on the following pages. They are arranged chronologically by date of construction.

THE POST-1940 HIGHWAY BRIDGES

Bridge Name	Bridge Number	County	Type	Date
Little River (Cavitt Creek)	19C18	Douglas	Timber Through Truss (Howe) Covered Bridge	1943
Santiam River	2541 and 8123	Marion-Linn	Twin Steel Through Tied Arches	1946 and 1958
Crabtree Creek (Bohemian Hall)	12890	Linn	Timber Through Truss (Howe) Covered Bridge	1947
Thomas Creek	8459	Curry	Steel Deck Truss (Warren with Verticals)	1961
Crooked River (Lake Billy Chinook)	16C06	Jefferson	Steel Suspension	1962
Columbia River (Astoria)	7949	Clatsop	Steel Through Truss (Cantilever)	1966
Thomas Creek (Shimanek)	12965	Linn	Timber Through Truss (Howe) Covered Bridge	1966
Willamette River (Fremont)	2529	Multnomah	Steel Half-Through Tied Arch	1973

Little River (Cavitt Creek) Bridge
Structure Number 19C18
Constructed - 1943
Timber Through Truss (Howe)
Covered Bridge
Cavitt Creek County Road 17
Glide vicinity, Douglas County
Ownership - Douglas County

This covered bridge is distinctive for its Tudor-shaped portal arches and unhewn timbers (as upper and lower chord members). Built by veteran Douglas County covered bridge builder Floyd C. Frear, the structure is a 70-foot housed Howe truss. The structure spans Little River near its confluence with Cavitt Creek, hence the name of the bridge. This covered bridge was included in the thematic nomination of Oregon's covered bridges to the National Register in 1979, but was not listed at the request of Douglas County.

Santiam River Bridges
Structure Numbers 2541 and 8123
Constructed - 1946 and 1958
Twin Steel Through Tied Arches
Pacific Highway 1 (I-5), M.P. 240.66
Jefferson vicinity, Marion-Linn counties
Ownership - State of Oregon

The easternmost of these two arch bridges was constructed in 1946 on the Pacific Highway, while its twin was built in 1958 when Interstate 5 was constructed. Both structures are three-span (180', 240' and 180') steel through tied arches flanked by six reinforced concrete deck girder approach spans. The 1946 structure serves northbound traffic, and the 1958 structure, southbound. Glenn S. Paxson, State Bridge Engineer, was the designer of the first bridge.

Crabtree Creek (Bohemian Hall) Bridge *
(Also, Richardson Gap Bridge)
Structure Number 12890
Constructed - 1947
Timber Through Truss (Howe)
Covered Bridge
Richardson Gap County Road 25 *
Scio vicinity, Linn County
Ownership - Linn County

One of three covered bridges in Oregon enclosed entirely with metal instead of wood siding, this structure is a 120-foot housed Howe truss. The bridge was known as the Richardson Gap Bridge after the Richardson family, who settled in the area in the 1880s. The bridge is also known as the Bohemian Hall Bridge from the nearby fraternal lodge established by Czechoslovakian immigrants in the 1920s. The bridge contains the open truss plan typical of seven Linn County covered bridges. The bridge was built by the Lindstrom Brothers at a cost of $40,000 in 1947. This covered bridge was included in the thematic nomination of Oregon's covered bridges to the National Register in 1979, but was not listed at the request of Linn County.

* *The Bohemian Hall covered bridge was dismantled in 1988 and placed in storage.*

Thomas Creek Bridge
Structure Number 8459
Constructed - 1961
Steel Deck Truss (Warren with Verticals)
Oregon Coast Highway 9, M.P. 374.78
Brookings vicinity, Curry County
Ownership - State of Oregon

This structure is the highest bridge in Oregon at 345 feet and spans a deep ravine on the Oregon Coast. (The second highest bridge is the High Bridge across the Crooked River in Jefferson County at 295 feet.) The bridge consists of three steel deck trusses (maximum span, 371 feet), supported on steel frame towers on concrete piers. The trusses are Warrens with verticals. The overall length of the bridge is 956 feet. The bridge was designed by Ivan D. Merchant, State Bridge Engineer, and constructed by the State Highway Department.

Crooked River (Lake Billy Chinook) Bridge
Structure Number 16C06
Constructed - 1962
Steel Suspension
Jordan County Road 579
Madras vicinity, Jefferson County
Ownership - Jefferson County

Located at the south end of Cove Palisades State Park in Jefferson County, this structure is a 588-foot steel suspension span. This bridge, another that crosses the Deschutes nearby, and the St. John's Bridge in Portland are the only large suspension bridges in Oregon. The bridge was built in conjunction with the Round Butte Dam and Lake Billy Chinook reservoir project. (Billy Chinook was an Indian scout and guide with the John C. Fremont and Kit Carson expedition to Central Oregon in 1843-44.)

Columbia River (Astoria) Bridge
Structure Number 7949
Constructed - 1966
Steel Through Truss (Cantilever)
Oregon Coast Highway 9, M.P. 0.00
Astoria, Clatsop County (Oregon) and
Megler vicinity, Pacific County (Washington)
Ownership - State of Oregon

At just over four miles in length (21,474 feet), the Columbia River Bridge at Astoria is the longest bridge in Oregon. The main span of the structure is a 2,468-foot steel cantilever through truss made up of two 618-foot outer truss sections and a 1,232-foot central truss span. This cantilever truss is flanked by five steel deck trusses, one-hundred-forty 80-foot concrete deck girder spans, and, at the Washington end of the bridge, seven 350-foot steel through truss spans. The bridge was designed jointly by the state highway departments of Oregon and Washington. Ivan D. Merchant was the state bridge engineer for Oregon. Construction began on the structure in 1962 and was completed in July 1966.

Thomas Creek (Shimanek) Bridge
Structure Number 12965
Constructed - 1966
Timber Through Truss (Howe)
Covered Bridge
Richardson Gap County Road 673
Scio vicinity, Linn County
Ownership - Linn County

One of the youngest of Oregon's covered bridges, this is the fifth covered bridge at this location. The earliest bridge at this site, an uncovered structure, was built in 1861. Later, covered bridges were constructed in 1891, 1904, 1921, and 1927. Named for the pioneer Shimanek family, this structure is a 130-foot housed Howe truss. The bridge was designed by W.A. Palmateer and built by the Hamilton Construction Company of Springfield. This covered bridge was not listed on the National Register at the request of Linn County, but was included in the thematic nomination of Oregon's covered bridges in 1979. It was finally added to the National Register in February 1987.

Willamette River (Fremont) Bridge
Structure Number 2529
Constructed - 1973
Steel Half-Through Tied Arch
Stadium Freeway 61 (I-405), M.P. 3.32
Portland, Multnomah County
Ownership - State of Oregon

When constructed, the Fremont Bridge was the largest of its type, a stiffened steel tied arch with an orthotropic upper deck. The 902-foot tied arch midspan was constructed off-site, floated into place, and its 6,000-ton weight hydraulically lifted 170 feet into position, establishing a place in the *Guinness Book of World Records* as the biggest lift ever made. The total length of the superstructure is 2,159 feet or 7,312 feet including approaches. The Fremont has no inwater pier supports, as the design keeps deadweight to a minimum. The bridge was designed for ten lanes of traffic, five on each level of the roadway. The Fremont Bridge was designed by Parsons, Brinckerhoff, Quade, and Douglas of New York, under contract by the Oregon State Highway Division. Ivan D. Merchant and Walter J. Hart were the state bridge engineers during the bridge planning and construction period.

APPENDIX G

MASTER LIST OF THE INVENTORIED BRIDGES

About 640 bridges were either field or plans inventoried for the historic highway bridges study.

These bridges are listed by county and alphabetically below in summary format. Highway bridges on county lines were arbitrarily assigned to one of the counties.

Legend

Burnt River - *Bridge Name*
707 - *Bridge Number*
T - *Structural type of the main or central span:*
- *Truss (T)*
- *Arch (A)*
- *Suspension (S)*
- *Moveable (M)*
- *Slab, Beam and Girder (BG)*

1922 - *Date of Completion*
f - *Inventory status:*
- *Field inventory (f)*
- *Plans inventory only (p)*

NR - *National Register eligible or listed*

BAKER COUNTY--13 Bridges

Burnt River, #700, T, 1922, f
Burnt River (Hereford Loop Rd.), #1C18, T, 1921, f
Eagle Creek, #1122A, T, 1925, f
North Powder River (Bidwell Lane), #661, T, 1937, f
North Powder Viaduct, BG, 1921, f
Pine Creek (Crow Lane), #1C826, T, 1911, f
Pine Creek (Dead Cow Butte), #1C834, T, 1911, f
Powder River (Beaver Creek Road), #1C202, T, 1911, f
Powder River (McCarty), #1C007, T, 1910, f
Powder River (Miles), #2314, T, 1937, f
Powder River (Myrtle Street), #4TO4, T, 1920, f
Powder River (Private), T, f
Pritchard Creek, #741, T, 1922, f

BENTON COUNTY--14 Bridges

Alsea River (Hayden) Covered Bridge, #14538, T, 1918, f, NR
Jackson Creek, #420A, BG, 1919, f
Lake Slough, #394, BG, 1919, f
Locke Creek, #419A, BG, 1919, f
Marys River, T, f

Marys River (Corvallis), #706, T, 1933, f
Marys River (Harris) Covered Bridge, #1441, T, 1936, f, NR
Marys River (Harris Road), #14478, T, 1914, f
North Fork Alsea River, #1204, T, 1927, f
North Fork Alsea River (Alsea), #2305, T, 1937, f
Oak Creek, #7T24, BG, 1940, f
Rock Creek, #1259, BG, 1926, f
Willamette River (Van Buren Street), #2728, M, 1913, f
Willamette Slough (Irish Bend) Covered Bridge, #45005-17, T, 1954, f, NR

CLACKAMAS COUNTY--23 Bridges

Abernethy Creek, #6218, BG, 1932, f
Abernethy Creek, #6223, T, 1915, f
Abernethy Creek (Holly Lane), #6217, T, 1933, f
Bear Creek (Faubion Road), #5C03, BG, 1921, f
Bull Run River #6571, T, 1894, f, NR
Clackamas River (Barton), #6560, A, 1936, f
Clackamas River (Carver), #1446, T, 1930, f

Clackamas River (Estacada), #2208, A, 1936, f
Clackamas River (McLoughlin), #1617, A, 1933, f, NR
Clackamas River (Park Place), #604, T, 1921, f
Eagle Creek (Doughty Road), #6561, T, 1930, f
Johnson Creek, #1137, BG, 1926, f
Molalla River, #1515, T, 1930, f
Molalla River, #2061, T, 1936, f
Oswego Creek, #6072, BG, 1934, f
Oswego Creek, #409, A, 1920, f, NR
Partial Viaduct, #2732, BG, 1940, f
Pudding River, #1559, BG, 1931, f
Salmon River (East Brightwood Loop), #1438, T, 1929, f
Sandy River (Lusted Road), #6580, T, 1894, f, NR
Sandy River (Sleepy Hollow), #1659, T, Ca. 1910, f
Tualatin River (Stafford Road), #2567, T, 1939, f
Willamette River (Oregon City), #357, A, 1922, NR

CLATSOP COUNTY--19 Bridges

Arch Cape Creek, #1797, BG, 1937, f
Austin's Point Half-Viaduct, #1878, BG, 1933, f

Beltline Overcrossing, #2418, BG, 1921, f
Beneke Creek, #1991, BG, 1934, p
Columbia River (Astoria) Bridge, #7949, T, 1966, f
East Fork Humbug Creek, #1832, BG, 1934, p
John Day River, #1827, M, 1933, f
Lewis and Clark River, #711, M, 1924, f
Neacoxie Creek (Del Ray Beach), #7C15, BG, 1930, p
Neawanna Creek, #1305, BG, 1930, f
Necanicum Creek (Skiberene), #1481, BG, 1930, f
Necanicum River (Seaside), #7C11, A, 1924, f, NR
Nehalem River, #2165, A, 1939, f
North Fork Quartz Creek, #2164, BG, 1939, f
Old Skipanon River, #1400, BG, 1929, f
Old Young's Bay Bridge, #330, M, 1921, f
Ravine, #7C17, BG, 1924, p
Soapstone Creek, #1319, A, 1928, f
Walluski River, #2320, M, 1938, p

COLUMBIA COUNTY--38 Bridges

Beaver Creek, #136, BG, 1918, f
Beaver Creek, #138, BG, 1918, f
Beaver Creek, #140, BG, 1918, f
Beaver Creek, #142, BG, 1918, f
Beaver Creek, #144, BG, 1918, f
Beaver Creek, #155, BG, 1918, f
Beaver Creek, #157, BG, 1918, f
Beaver Creek, #9C57, BG, 1918, f
Beaver Creek, #9C59, BG, 1918, f
Beaver Creek, #13611A, BG, 1918, f
Beaver Creek (Kukkla Road), #9C83, T, Ca. 1915, f
Beaver Slough (Inglis), #1574, BG, 1931, p
Clatskanie River, #8403, T, f
Columbia River (Lewis and Clark), #433/1, T, 1930, f, NR
Fishhawk Creek (Birkenfeld), #9C118, T, 1935, f
Honeyman Creek, #2667, BG, Ca. 1920, p
McNulty Creek, #2599, BG, Ca. 1920, p
Merrill Creek, #13363, BG, Ca. 1920, p
Milton Creek, #303, BG, 1920, f
Milton Creek, Lower, (McDonald), #13757, T, 1914, f, NR
Nehalem River, #1415, BG, 1929, p
Nehalem River (Mile), #2323, T, 1938, f
Nehalem River (Natal), #13665A, T, 1935, f
North Fork Scappoose Creek, #2668, BG, Ca. 1920, p
North Fork Scappoose Creek, #13743, BG, Ca. 1920, p
North Fork Scappoose Creek, #13352A, BG, Ca. 1920, p
North Fork Scappoose Creek, #13762A, BG, Ca. 1920, p
North Fork Wolf Creek, #2028A, BG, 1938, p
Rock Creek, #1508, BG, 1930, p
South Beaver Creek, #9C158, BG, 1918, f
South Fork Scappoose Creek, #2670, BG, Ca. 1920, p
South Fork Scappoose Creek, #13751, BG, Ca. 1920, p
Tide Creek, #338, BG, 1920, f
Tide Creek, #13626A, BG, Ca. 1920, p
Unnamed Creek, #13610A, BG, Ca. 1920, p
West Creek #13774, BG, Ca. 1920, p
Westport Slough, BG, f
Westport Slough (Midland-Marshfield), #13825, T, 1936, f

COOS COUNTY--10 Bridges

Catching Slough, #2278C, T, 1939, f
Coos Bay (McCullough Memorial), #1823, T, 1936, f, NR
Coquille River, #598, T, 1922, f, NR
Isthmus Slough (Eastside), #1132F, M, 1931, f
Reservoir Creek, #1060, BG, 1924, p
Sandy Creek (Remote) Covered Bridge, #4037, T, 1921, f, NR
South Fork Coquille River, #1942A, T, 1934, f
South Slough, #1940, T, 1934, f
Southern Pacific Railroad Undercrossing, #1950, BG, 1935, f
Tenmile Creek, #949, BG, 1924, f

CROOK COUNTY--6 Bridges

Bear Creek, #7C03, T, 1928, f
Crooked River, #16132, T, 1923, f
Crooked River (Elliott Lane), #7C12, T, Ca. 1914, f
Irrigation Ditch, #13C22, BG, 1940, p
McKay Creek, #13C21, BG, 1938, p
Ochoco Creek, #2201, BG, 1920, p

CURRY COUNTY--7 Bridges

Euchre Creek, #15C31, BG, 1927, f
Hunter Creek, #15C10, BG, 1928, f
Myer's Creek, #995, BG, 1924, f
Myrtle Creek, #15C15, BG, 1925, p
Private Road, #15C28, BG, Ca. 1930, p
Rogue River (Gold Beach), #1172, A, 1931, f, NR
Thomas Creek, #8459, T, 1961, f

DESCHUTES COUNTY--4 Bridges

Oregon Trunk Railroad Undercrossing, #3365, BG, 1931, p
Skyliners Road (Brook's Log Railroad), #9C63, BG, 1920, p
Swalley Canal (Rock O' the Range) Covered Bridge, BG, 1963, f, NR
Tumalo Irrigation Ditch, #17C02, A, 1914, f, NR

DOUGLAS COUNTY--35 Bridges

Calapooya Creek, #4655, BG, p
Calapooya Creek (Oakland), #603, T, 1925, f
Calapooya River (Rochester) Covered Bridge, #1893, T, 1933, f, NR
County Road 368, #368, BG, p
Deer Creek, #26TO3, BG, 1919, f
Deer Creek, #26TO6, T, 1928, f
Diamond Creek, #2079, T, 1935, f
Elk Creek (First Crossing), #1614, T, 1931, f
Elk Creek (Second Crossing), #1601, BG, 1931, f
Elk Creek (Third Crossing), #1465, T, 1931, f
Elk Creek (Fourth Crossing), #1406, T, 1931, f
Elk Creek (Roaring Camp) Covered Bridge, T, 1929, f, NR
Little River (Cavitt Creek) Covered Bridge, #19C18, T, 1943, f
Little River, T, Ca. 1910, f
North Umpqua River (Mott), #4712-000-0.1, A, 1936, f, NR

North Umpqua River (Winchester), #839, A, 1924, f, NR
Old Vandine Creek Connnection Route, #234, BG, 1918, f
Paradise Creek, #1697, BG, 1932, p
Pass Creek Covered Bridge, #19B01, T, 1925, f, NR
Smith River, #2129, BG, 1936, f
South Myrtle Creek (Neal Creek) Covered Bridge, #10C220, T, 1939, f, NR
South Umpqua River (Myrtle Creek), #490A, A, 1922, f, NR
South Umpqua River, #1192, T, 1926, f
South Umpqua River, #1586, T, 1932, f
South Umpqua River, #2556, T, 1939, f
South Umpqua River (Worthington), #7408, T, Ca. 1910, f
South Umpqua River (Milo Academy) Covered Bridge, BG, 1962, f, NR
South Umpqua River (Winston), #1923, A, 1934, f
Southern Pacific Railroad Overcrossing, #9533, BG, p
Sutherlin Creek, #431, BG, 1921, p
Tahkenitch Creek, #1602, BG, 1929, f
Umpqua River (Scottsburg), #1318, T, 1929, f
Umpqua River (Reedsport), #1822, M, 1936, f, NR
Umpqua River (Bullock), #10C358, T, Ca. 1910, f
Yellow Creek, #1995, BG, 1934, p

GILLIAM COUNTY--1 Bridge

Rock Creek (Olex), T, Ca. 1905, f, NR

GRANT COUNTY--4 Bridges

Branson Creek, #240A, BG, 1939, p
Dixie Creek, #2466, BG, 1939, p
Holmes Creek, #242, BG, 1937, p
Middle Fork John Day River, #1922, T, 1934, f

HARNEY COUNTY--None

HOOD RIVER COUNTY--12 Bridges

Columbia River (Bridge of the Gods), T, 1926, f, NR
Columbia River (White Salmon), #6645, M, 1924, f
East Fork Hood River, #640, T, 1922, f
East Fork Hood River, #1939, T, 1934, f
Gorton Creek, #27C35, BG, 1918, f, NR
Herman Creek, #159, BG, 1918, f
Hood River (Hood River), #200, A, 1918, f
Hood River (Tucker), #1600, A, 1932, f
Rock Slide Viaduct, #504, BG, 1920, f, NR
Ruckel Creek, BG, 1917, f, NR
Ruthton Point Viaduct, #273, BG, 1918, f, NR
South Fork Hood River (Sahalie Falls), #3545-000-1.2, A, 1928, f, NR

JACKSON COUNTY--24 Bridges

Antelope Creek Covered Bridge, #29C202, T, 1922, f, NR
Applegate River, #1992, T, 1934, f
Applegate River (McKee) Covered Bridge, #29C471, T, 1917, f, NR
Bear Creek, #19T31, BG, 1912, f
Big Butte Creek, #29C233, T, 1912, f
Birdseye Creek, #412A, BG, 1920, f
Bybee Creek, #3460, BG, 1932, p
Creek (Trail), T, Ca. 1925, f
Dollarhide Overcrossing, #3781, BG, 1914, f, NR
Elk Creek, #830, T, 1923, f
Evans Creek, #9089, A, 1914, f
Evans Creek (Wimer) Covered Bridge, #29C211, T, 1927, f, NR
Jackson Creek, #29C105, BG, 1939, p
Jenny Creek, #660, BG, 1920, f
Kenne Creek, #596, BG, 1940, p
Lost Creek Covered Bridge, #29C262, T, 1919, f, NR
Miller's Gulch, #413, BG, 1920, f
Neil Creek, #380, BG, 1920, f
North Fork Rogue River (Prospect Arch), #29C281, A, 1923, f
Rogue River (Rock Point), #332A, A, 1920, f, NR
Rogue River (Gold Hill), #576, A, 1927, f, NR
Sam's Creek (Pelton Lane), T, Ca. 1925, f
Sardine Creek, #1937, BG, 1937, p
Steinman Overcrossing, #3780, BG, 1914, f, NR

JEFFERSON COUNTY--4 Bridges

Crooked River (High), #600, A, 1926, f, NR
Crooked River (Lake Billy Chinook), #16C06, S, 1962, f
Deschutes River, #1910, BG, 1934, p
Lake Creek, #3785, BG, 1930, p

JOSEPHINE COUNTY--16 Bridges

Anderson Creek, #1194A, BG, 1927, p
Applegate River, #1332, T, 1927, f
Applegate River, #1985, T, 1935, f
Deer Creek, #517010, T, 1921, f
Democrat Creek, #584030, BG, 1925, p
Grave Creek Covered Bridge, #141005, T, 1920, f, NR
Illinois River, #531515, T, 1926, f
Jack's Creek, #33C30, BG, 1940, p
Rogue River (Caveman), #1418, A, 1931, f, NR
Rogue River (Robertson), #1592, T, Ca. 1909, f
Slate Creek, #1269A, BG, 1926, p
Southern Pacific Railroad (Wolf Creek) Overcrossing, #33C36, BG, 1920, f
Williams Creek, #2379, A, 1917, f, NR
Williams Creek, #406005, T, 1935, f
Williams Creek, #450015, BG, 1930, p
Wolf Creek, #114005, BG, 1921, f

KLAMATH COUNTY--12 Bridges

A Canal (11th St. Bridge), #62002, BG, 1938, f
Irrigation Canal, #18C001, BG, 1940, p
Irrigation Canal (U.S.R.S.), #3881, BG, 1930, p
Irrigation Canal (U.S.R.S.), #1419A, BG, 1930, p
Klamath River (Keno), #986, A, 1931, f
Klamath Straits, #1794, BG, 1933, p
Link River, #1579, BG, 1931, f
Link River (Fremont), #63001, BG, 1926, f
Lost River, #35C213, T, 1930, f
Oregon and California Railroad Overcrossing, #2147, BG, 1936, p
Southern Pacific Railroad Overcrossing, #2020, BG, 1936, p

Southern Pacific Railroad Overcrossing, #2474, BG, 1940, p

LAKE COUNTY--5 Bridges

Goose Lake Swale, #2345, BG, 1940, p
Hart Lake Canal #2, #37C03, BG, 1937, p
Irrigation Canal, #2358, BG, 1940, p
Silver Creek, #3908, BG, 1940, p
South Branch Buck Creek, #3907, BG, 1940, p

LANE COUNTY--62 Bridges

Amazon Channel, #40011, BG, 1919, f
Amazon Channel, #40042, BG, 1930, p
Amazon Creek, #1755H, BG, 1932, p
Approach Road, #16003, BG, 1931, f
Big Creek, #1180, A, 1931, f
Bob Creek, #1177, BG, 1931, f
Cape Creek, #1113, A, 1931, f, NR
Coyote Creek (Battle Creek) Covered Bridge, #39C409, T, 1922, f, NR
Creek, #1755C, BG, 1932, p
Creek, #1755N, BG, 1932, p
Creek, #1755P, BG, 1932, p
Creek, #1805C, BG, 1933, p
Creek, #2229C, BG, 1937, p
Creek, #3946, BG, 1919, f
Creek, #39C362, BG, 1919, f
Creek, #39C301, BG, 1919, p
Creek, #39C302, BG, 1919, f
Creek, #9C363, 1931, f
Creek, (River Road), #3948, BG, 1919, f
Crow Creek, #39C355, BG, 1931, f
Cummins Creek, #1182, A, 1931, f
Deadwood Creek, #1331, T, 1928, f
Deadwood Creek Covered Bridge, #16-9W-25, T, 1932, f, NR
Fall Creek (Pengra) Covered Bridge, #18-1W-32, T, 1938, f, NR
Fall Creek (Unity) Covered Bridge, #14721, T, 1936, f, NR
Flat Creek, #1805B, BG, 1933, p
Flat Creek, #3952, BG, 1915, f
Half-Viaduct, #7185, BG, 1939, f
Half-Viaduct, #7186, BG, 1939, f
Half-Viaduct, #7187, BG, 1939, f
Half-Viaduct, #7188, BG, 1939, f
Horse Creek Covered Bridge, #16-5E-24, T, 1930, f, NR
Lake Creek (Nelson Mountain) Covered Bridge, #39C386, T, 1928, f, NR
Lingo Slough, #1755D, BG, 1932, p
Lost Creek (Parvin) Covered Bridge, #19-1W-21, T, 1921, f, NR
McKenzie River (Armitage), #1336, T, 1927, f
McKenzie River (Belknap) Covered Bridge, #39C123, T, 1966, f, NR
McKenzie River (Goodpasture) Covered Bridge, #39C118, 1939, f, NR
Middle Fork Willamette River (Lowell) Covered Bridge, #19-1W-23, T, 1945, f, NR
Mill Creek (Wendling) Covered Bridge, #39C174, T, 1938, f, NR
Mill Race, #1, BG, 1914, f
Mohawk River (Ernest) Covered Bridge, #39C176, T, 1938, f, NR
Mosby Creek Covered Bridge, #39C241, T, 1920, f, NR
Mosby Creek (Stewart) Covered Bridge, #39C243, T, 1930, f, NR
North Fork of the Middle Fork Willamette River (Office) Covered Bridge, T, 1944, f, NR
Overcrossing (Highway 62), #6663, BG, 1933, p
Overcrossing (Highway 225 and Southern Pacific Railroad), #3953, BG, 1917, f
Overcrossing (Southern Pacific Railroad), #2138, BG, 1937, p
Rock Creek, #1176, BG, 1931, f
Row River (Currin) Covered Bridge, #39C242, T, 1925, f, NR
Row River (Dorena) Covered Bridge, #21-2W-24A, T, 1949, f, NR
Salmon Creek, #2073, T, 1933, f
Siltcoos River, #982, BG, 1930, f
Siuslaw River (Florence), #1821E, M, 1936, f, NR
Siuslaw River (Richardson), #39C501, T, 1912, f, NR
Spring Creek, #39C303, BG, 1919, f
Sutton Creek, #1494, BG, 1930, f
Tenmile Creek, #1181, A, 1931, f
Wildcat Creek Covered Bridge, #39C446, T, 1925, f, NR
Willamette River, #583, T, 1925, f
Willamette River (Springfield), #1223, T, 1929, f, NR
Willamette River, #1626, T, 1932, f

LINCOLN COUNTY--14 Bridges

Alsea Bay (Waldport), #1746, A, 1936, f, NR
Cook's Chasm, #1174, BG, 1931, f
Depoe Bay, #2459, A, 1927, f, NR
Drift Creek Covered Bridge, #10803, T, 1914, f, NR
Five Rivers (Fisher School) Covered Bridge, #90101, T, 1919, f, NR
Lint Creek, #4166, BG, 1931, p
North Fork Yachats River Covered Bridge, #12037, T, 1938, f, NR
Rocky Creek (Ben Jones), #1089, A, 1927, f, NR
Salmon River, #4192, A, 1930, f
Yaquina Bay (Newport), #1820, A, 1936, f, NR
Yaquina River, #683, T, 1923, f
Yaquina River, #1401, T, 1929, f
Yaquina River, #1402, T, 1929, f
Yaquina River (Chitwood) Covered Bridge, #41C09, T, 1926, f, NR

LINN COUNTY--27 Bridges

Albany Canal, #43C12, BG, 1930, p
Calapooia River, #2373, T, 1940, f
Calapooia River, #12569, T, 1940, f
Calapooia River (Crawfordsville) Covered Bridge, #12819, T, 1932, f, NR
Calapooia River (Matlock), #2033, T, 1934, f
Cox Creek, #2515A, BG, 1940, p
Crabtree Creek (Bohemian Hall) Covered Bridge, #12890, T, 1947, f
Crabtree Creek (Hoffman) Covered Bridge, #1724, T, 1936, f, NR
Crabtree Creek (Larwood) Covered Bridge, #12876, T, 1939, f, NR
Lake Creek, #738, BG, 1922, p
Mill Creek, #12723, BG, 1938, p
North Santiam River, #2454, BG, 1938, p
Oak Creek, #737, BG, 1922, p
Pamela Creek, #6805, BG, 1934, p
Santiam River (Cascadia Park), #1356, T, 1928, f, NR
South Fork Santiam River (Short) Covered Bridge, #14025, T, 1945, f, NR
South Santiam River, T, f
South Santiam River (Garland), #1577, T, 1931, f
Southern Pacific Railroad Overcrossing, #2380, BG, 1940, p
Thomas Creek, #2346, BG, 1940, p
Thomas Creek, #2623, T. 1939, f
Thomas Creek (Gilkey) Covered Bridge, #12943, T, 1939, f, NR
Thomas Creek (Hannah) Covered Bridge, #12948, T, 1936, f, NR

Thomas Creek (Jordan) Covered Bridge, #12958, T, 1937, f, NR
Thomas Creek (Shimanek) Covered Bridge, #12965, T, 1966, f
Thomas Creek (Weddle) Covered Bridge, #12935, T, 1937, f, NR
Willamette River (Albany), #1025D, T, 1925, f, NR

MALHEUR COUNTY--22 Bridges

Antelope Canal, #1947, BG, 1937, p
Cow Canal, #45C607, BG, 1913, p
Cow Creek, #45C609, T, Ca. 1910, f, NR
Dead Ox Pump Canal (Pioneer), #45C133, BG, 1935, p
Faulk Island, #45C301, T, 1915, f
Gwynn Undercrossing (Union Pacific Railroad), #4340, BG, 1913, p
Irrigation Canal, #19727, BG, 1940, p
Jonesboro Undercrossing (Union Pacific Railroad), #4344, BG, 1913, p
Jordan Creek, #1948, T, 1937, f
Jordan Creek, #45C53, T, 1915, f
Jordan Creek Overflow, #2430, BG, 1937, p
Jordan Creek Overflow, #45C601, T, 1910, f
Malheur River #559A, T, 1921, f
Malheur River, #45C141, T, 1912, f
Malheur River (Red), #45C403, 1926, f
Malheur River (Speedy), #1551, BG, 1931, p
North Fork Malheur River, #943, T, 1923, f
Owyhee Canal, #34C214, BG, 1934, p
Owyhee River, #1945, T, 1937, f
Owyhee River, #45C611, T, 1906, f, NR
Riverside Canal, #45C133, BG, 1935, p
Snake River (Ontario Spur), #1000, T, 1925, f

MARION COUNTY--31 Bridges

Abiqua Creek (Gallon House) Covered Bridge, #5381, T, 1917, f, NR
Abiqua Creek, T, f
Abiqua Creek, #1260, T, 1926, f
Abiqua Creek, #47C63, T, 1926, f
Croisan Creek, #701, BG, 1940, f
Drift Creek, #1534, BG, 1930, f
Mill Creek, BG, f
Mill Creek, BG, 1938, f
Mill Creek, #1502, BG, 1930, f
Mill Creek, #1427, BG, 1929, f
Mill Creek, #2021, BG, 1934, f
Mill Creek, #6664, BG, 1929, f
Mill Creek (Summer Street, N.E.), #13575, BG, 1929, f
Mill Creek (Front Street, N.E.), #138114, A, 1913, f, N R
Mill Creek, #14834, BG, 1914, f
Mill Creek, #308106, BG, 1937, f
Portland Road, N.E., Undercrossing (Southern Pacific Railroad), #2131, BG, 1936, f
Pringle Creek, BG, f
Pringle Creek (Commercial Street, South), #1340, BG, 1928, f
Pringle Creek (Liberty Street S.E.), #1357, BG, 1928, f, NR
Pringle Creek/Shelton Creek (Church Street, S.E.), #608, BG, 1929, f, NR
Pudding River, #1917, BG, 1930, f
Pudding River Relief Channel, #1830, BG, 1922, p
Santiam River (Jefferson), #1582, A, 1933, f, NR
Santiam River, #2058, T, 1934, f
Santiam River, #2541, A, 1946, f
Santiam River, #8123, A, 1958, f
Shelton Ditch, #405, BG, 1928, f
Shelton Ditch, #148136, A, 1940, f
West Fork Mill Creek, #2272, BG, 1937, p
Willamette River (Center Street), #123A, T, 1918, f

MORROW COUNTY--19 Bridges

Blackhorse Creek, #49C15, BG, 1920, f
Brandy Creek, #670A, BG, 1920, f
Butter Creek (Pine City), BG, #10928, 1928, p
Creek, #49C08, BG, 1920, f
East Fork Dry Creek, #5638, 1940, p
Fuller Canyon, #49C02, BG, 1915, f
Reitner Creek, #672, BG, 1920, f
Rhea Creek (Spring Hollow), #49C05, T, 1909, f, NR
Rhea Creek, #49C23, T, 1916, f
Rhea Creek (Snyder), #10891, BG, 1928, p
Rock Creek, #49C16, BG, 1920, f
Shobe Canyon, #4497, BG, 1920, f
West Fork Dry Creek, #5637, 1940, p
Willow Creek, T, f
Willow Creek (Cecil), #10949, T, 1909, f, NR
Willow Creek, #10958, T, 1938, f
Willow Creek, #49C01, BG, 1920, f
Willow Creek (Clark's Canyon), #49C09, BG, 1920, f
Willow Creek (Court Street), #778A, T, 1921, f

MULTNOMAH COUNTY--74 Bridges

Balch Gulch, #25B15, T, 1905, f, NR
Beaver Creek (Sandy River Overflow), #4522, BG, 1912, f, NR
Bridal Veil Falls, #823, BG, 1914, f, NR
Columbia River (Interstate Northbound), #1377A, T, 1917, f, NR
Columbia River (Interstate Southbound), #7333, M, 1958, f
Columbia Slough (N.E. Union Avenue), #1377C, BG, 1916, f
Crown Point Viaduct, #4524, BG, 1914, f, NR
Draw (N.W. Cornell Road), #11040, BG, 1940, p
Eagle Creek, A, 1915, f, NR
East Multnomah Falls Viaduct, #841, BG, 1914, f, NR
Gully (N.E. Glisan St.), #25B33, BG, 1911, f
Half-Viaduct (Lower Columbia River Highway), #5291, BG, 1916, f
Horsetail Falls, #4543, BG, 1914, f, NR
Johnson Creek (S.E. Cathey Road), #25TO4, BG, 1927, p
Johnson Creek (S.E. Foster Road), #11086, BG, 1915, f
Johnson Creek (S.E. Tacoma Street), #51C02, BG, 1915, f
Johnson Creek (S.E. Umatilla Street), #51C02, BG, 1915, f
Latourell Creek, #4527, A, 1914, f, NR
McCarthy Creek, BG, Ca. 1914, f
Moffet Creek, #2194, A, 1915, f, NR
Multnomah Creek, #4534, A, 1914, f, NR
N.E. 12th Avenue Overcrossing (Banfield Freeway and Union Pacific Railroad), #7039, BG, 1910, f
N.E. 21st Avenue Overcrossing (Banfield Freeway and Union Pacific Railroad), #7035, BG, 1912, f
N.E. 28th Avenue Overcrossing (Banfield Freeway and Union Pacific Railroad), #7034, BG, 1908, f
N.E. 42nd Avenue Overcrossing (Portland Highway), #2485, BG, 1939, p

N.E. 47th Avenue Overcrossing (Banfield Freeway and Union Pacific Railroad), #7024, BG, 1917, f
N.E. 53rd Avenue Overcrossing (Banfield Freeway and Union Pacific Railroad), #7027, BG, 1917, f
N.E. 60th Avenue Overcrossing (Banfield Freeway and Union Pacific Railroad), #7028, BG, 1917, f
N.E. 74th Avenue Overcrossing (Banfield Freeway), #7030, BG, 1917, f
N.E. Grand Avenue Overcrossing (Banfield Freeway and Union Pacific Railroad), #7040, BG, 1908, f
N.E. Halsey Street Overcrossing (Banfield Freeway), #A2142, BG, 1918, f
N.E. Halsey Street Overcrossing (Banfield Freeway), #7029, BG, 1918, f
N.E. Union Avenue Overcrossing (Banfield Freeway and Union Pacific Railroad), #2350A, BG, 1908, f
N.E. Union Avenue Overcrossing (Union Pacific Railroad at Baldwin Street), #5290, BG, 1916, f
N.W. Alexandra Avenue Viaduct, #25B14, A, 1922, f, NR
Newbury Street Viaduct, #1983, BG, 1934, p
North Columbia Boulevard Undercrossing (Spokane, Portland, and Seattle Railroad), #9685, BG, 1909, f
North Denver Avenue Overcrossing (North Schmeer Road Connection), #4517, BG, 1916, p
North Fessenden Street Overcrossing (Spokane, Portland, and Seattle Railroad), #6.7, T, 1909, f, NR
North Lombard Street Overcrossing (Spokane, Portland, and Seattle Railroad), #6.2, T, 1909, f
North Lombard Street Overcrossing (Union Pacific Railroad), #25B01, BG, 1930, f
North Willamette Boulevard Overcrossing (Spokane, Portland, and Seattle Railroad), #5.8, T, 1909, f, NR
Oneonta Gorge Creek (Old), #4542, BG, 1914, f, NR
Oneonta Gorge Creek (New), #7108A, BG, 1948, f, NR
Oregon Slough (Interstate 5), #1377B, BG, 1916, f
Riverside Canyon, #1185A, BG, 1926, p
Sandy River (Stark Street), #11112, T, 1914, f

Sandy River (Troutdale), #2019, T, 1912, f, NR
S.E. Bybee Boulevard Overcrossing (S.E. McLoughlin Boulevard and Southern Pacific Railroad), #2026 and #25B50-1, BG, 1934, f
S.E. Holgate Overcrossing (Southern Pacific Railroad), #25B44, BG, 1916, f
Stark Street Viaduct, #11113, A, 1915, f, NR
Steel Bridge Approach (Northbound), #2733E, BG, 1912, f
S.W. Barbur Boulevard Overcrossing (S.W. Multnomah Boulevard), #2010, BG, 1935, p
S.W. Bertha Boulevard Undercrossing (S.W. Capitol Highway), #24TO1, BG, 1915, f
S.W. Capitol Highway Overcrossing (S.W. Multnomah Boulevard), #25TO3, BG, 1927, f
S.W. Hawthorne Street Overcrossing (West End Hawthorne Bridge), #2757A, BG, 1940, p
S.W. Osage Street Viaduct, #25B35, BG, 1929, p
S.W. Terwilliger Boulevard Overcrossing (Interstate 5), #8392, T, 1914, f
S.W. Vista Avenue Viaduct, #25B36, A, 1926, f, NR
Tanner Creek, #2062, BG, 1915, f, NR
Toothrock and Eagle Creek viaducts, BG, 1915, f, NR
Union Pacific Railroad Overcrossing (Crown Point Highway, Troutdale), #4521, BG, 1912, f
Valley (S.W. Vista Avenue) Crossing, #25B41, BG, 1914, f
Wahkeena Falls, #4533, BG, 1914, f, NR
West Multnomah Falls Viaduct, #840, BG, 1914, f, NR
Willamette River (Broadway), #6757, M, 1913, f, NR
Willamette River (Burnside), #511, M, 1926, f, NR
Willamette River (Fremont), #2529, A, 1973, f
Willamette River (Hawthorne), #2757, M, 1910, f, NR
Willamette River (Ross Island), #5054, T, 1926, f, NR
Willamette River (Sellwood), #6879, T, 1925, f
Willamette River (Steel), #2733, M, 1912, f, NR

Willamette River (St. John's), #6497, S, 1931, f, NR
Young Creek (Shepperd's Dell), #4528, A, 1914, f, NR

POLK COUNTY--12 Bridges

Ash Creek, #4599, BG, 1916, f
Berry Creek, #871A, BG, 1923, p
Gooseneck Creek, #2015, BG, 1934, p
Luckiamute (Grant Road) River, #53C139, T, 1918, f
Luckiamute River, #53C122, T, 1928, f
Rickreall Creek, T, f
Rickreall Creek (Dallas), #231A, A, 1918, f
Rickreall Creek (Pumping Station) Covered Bridge, T, 1916, f, NR
Ritner Creek Covered Bridge, #1251, T, 1927, f, NR
South Yamhill River, #1344, T, 1928, f
South Yamhill River, #2081, BG, 1935, p
South Yamhill River (Steel), #53C05, T, Ca. 1910, f

SHERMAN COUNTY--3 Bridges

Finnegan Creek, #28C03, BG, 1920, p
Gurkin Canyon Creek, #1833, BG, 1933, p
Hay Creek, #28C02, BG, 1919, f

TILLAMOOK COUNTY--32 Bridges

Beaver Creek, #2762, BG, 1916, f
Beaver Creek, #4654, BG, 1916, f
Big Nestucca River (Condor), #555, T, 1921, f
Cedar Creek, #4673, BG, 1920, p
Chasm (Neahkahnie Mountain), #2723, BG, 1937, f, NR
Dougherty Slough, #1528, BG, 1931, f
Farmer Creek, #4659, BG, 1916, p
Half-Viaduct (Neahkahnie Mountain), #1951, BG, 1940, f
Half-Viaduct (Neahkahnie Mountain), #1954, BG, 1940, f
Half-Viaduct (Neahkahnie Mountain), #1955, BG, 1940, f
Hoquarten Slough, #1500, BG, 1920, p
Juno Overcrossing (Southern Pacific Railroad), #505, BG, 1931, f
Kilchis River, #455, T, 1920, f
Killam Creek, #57C29, A, 1914, f, NR

Lake Lytle Outlet, #2349, BG, 1938, f
Little Nestucca River, #1861, T, 1934, f
Miami River, #1226, T, 1926, f
Necarney Creek, #2311, BG, 1937, f, NR
Nehalem River, #574, M, 1921, f
Nehalem River, #1217, T, 1927, f
Nestucca River, #1372, T, 1931, f
North Fork Nehalem River (Scovill), #633, T, 1936, f
North Fork Trask River, #57C24, T, 1927, f
Short Sand Beach Creek, #2312, BG, 1937, f
Simmons Creek, #877, BG, 1922, p
Three Rivers, #4675, BG, 1920, p
Tiger Creek, #4651, BG, 1919, p
Trask River (Tone), #2470, T, 1935, f
West Beaver Creek, #2202, A, 1914, f
Wilson River, #1499, A, 1931, f, NR
Wilson River Overflow (Boquist Road), #57C60, BG, 1920, f
Wilson River Slough, #1498, BG, 1931, f

UMATILLA COUNTY--42 Bridges

A-Line Canal #475, BG, 1920, f
A-Line Canal, #10389, T, 1920, f
A-Line Canal, #10469, BG, 1920, f
Birch Creek, #10566, T, 1927, f
Camas Creek, #4729, T, 1932, f
Furnish Ditch, #10449, BG, 1935, p
Greasewood Creek, #59C590, T, 1907, f
Irrigation Canal, #385, BG, 1924, p
Maxwell Canal, #474A, BG, 1920, f
Maxwell Ditch, #10405, BG, 1935, f
McKay Creek (Sumac), #10624, T, 1935, f
Meacham Bridge, #447, BG, 1925, p
North Fork John Day River, #4729, T, 1932, f
Old Mill Race, #24T04, BG, 1914, f, NR
Pine Creek, #10712, T, Ca. 1910, f
Pine Creek, #59C528, T, Ca. 1910, f
Pine Creek, #59C534, T, Ca. 1910, f, NR
Pine Creek, #59C542, T, 1939, f
Pine Creek (Dry Creek), #59C535, T, 1935, f
Potts Canal, #10575, T, 1927, f
Stanfield Drain Ditch, #10488, T, 1939, f

Umatilla River (Umatilla), #624A, A, 1925, f
Umatilla River, #1360, T, 1928, f
Umatilla River, #10451, T, 1926, f
Umatilla River (S.E. 8th Street), #24T01, T, 1909, f, NR
Umatilla River, #24T02, BG, 1911, f
Umatilla River (S.W. 10th Street), #24T03, T, 1913, f, NR
Umatilla River (Riverside), #4697, T, 1916, f
Union Pacific Railroad Overcrossing (Umatilla), #1628, BG, 1933, p
Union Pacific Railroad Overcrossing (Riverside), #4696, BG, 1916, f
Union Pacific Railroad Undercrossing (S.E. Byers), #24TO7, BG, 1932, f
U.S. Feed Canal, #10451, T, 1926, f
U.S. Feed Canal, #10475, T, 1939, f
U.S.R.S. Feed Canal, #476, BG, 1920, f
U.S.R.S. Feed Canal, #478A, BG, 1920, f
U.S.R.S. Feed Canal, #479A, BG, 1920, f
Walla Walla River, #1350, T, 1928, f
Wild Horse Creek, #1033, T, 1924, f
Wild Horse Creek, #10539, T, 1929, f
Wild Horse Creek, #59C363, T, 1928, f
Wild Horse Creek (Adams), #10560, BG, 1924, p
Wild Horse Creek (McBride), #10708, BG, 1936, p

UNION COUNTY--15 Bridges

Catherine Creek (Lower Cove), #1157, T, 1930, f
Five Point Creek, #718, T, 1921, f
Gekler Slough, #61C34, BG, 1940, p
Grande Ronde River (McKennen Road), #61C19, T, 1925, f
Grande Ronde River (Old Rhinehart), #799, T, 1922, f, NR
Grande Ronde River (Perry Overcrossing), #626, A, 1924, f
Grande Ronde River (Rhinehart), #61C20, T, 1925, f
Grande Ronde River (Striker Lane), T, 1930, f
Grande Ronde River (Yarrington), #61C16, T, 1906, f
Little Creek, #61C27, BG, 1930, p
Little Creek, #61C29, BG, 1930, p

Red Pepper Slough, #61C41, T, 1930, f
State Ditch (Lower Cove), #1495, T, 1930, f
State Ditch (Ruckman), #61C14, T, 1925, f
Wright Slough (Conley Creek), #837A, BG, 1922, p

WALLOWA COUNTY--8 Bridges

Grande Ronde River (Troy), #32C62, T, 1910, f, NR
Imnaha River, T, f
Prairie Creek (River Street), #4878, A, 1931, f
Wallowa River (Stein), #63C38, T, 1030, f
Wallowa River (General Kenvill), #63C24, T, 1937, f
Wallowa River (Spur), #32C12A, T, 1920, f
Wallowa River (Water Canyon), T, f
Wallowa River (Wayne Wolfe), #32C25, T, 1920, f

WASCO COUNTY--21 Bridges

Chenoweth Creek, #506, BG, 1920, f, NR
Clear Creek, #2204, BG, 1932, p
Columbia River (The Dalles), #6635Q, T, 1954, f
Deschutes River (Maupin), #966, T, 1929, f
Draw, #65C44, 1940, p
Draw, #65C67, 1940, p
Dry Canyon Creek, #524, A, 1921, f, NR
Eightmile Creek, #106, BG, 1917, f
Eightmile Creek, #893, BG, 1922, f
Eightmile Creek, #895, BG, 1923, f
Eightmile Creek, #65C19, BG, 1915, p
Fifteenmile Creek (Adkisson), #1095, A, 1925, f, NR
Fifteenmile Creek (Fax), #15711, T, 1930, f
Fifteenmile Creek (Seufert) Viaduct, #308, BG, 1920, f, NR
Hog Creek Canyon (Rowena Dell), #523, BG, 1920, f, NR
Mill Creek (West Sixth), #464, BG, 1920, f, NR
Mosier Creek (State Road), #118, BG, 1917, f

Mosier Creek, #498, A, 1920, f, NR
Rock Creek, #65C63, BG, 1918, f, NR
Warm Springs River, #2619, BG, 1940, p
White River, #65C64, T, 1923, f

WASHINGTON COUNTY--8 Bridges

Gaston Slough, #67C05, BG, 1921, p
Nehalem River, #2364, BG, 1940, p
Scoggins Creek, #67C03, BG, 1921, p
Tualatin River, #1767, BG, 1931, p
Tualatin River, #14175, T, 1929, f
Tualatin River (Fern Hill Road), #1335, T, 1933, f
Tualatin River (Minter Bridge Road), #1575, BG, 1931, f
Wolf Creek, #2029, BG, 1938, p

WHEELER COUNTY--1 Bridge

Kahler Creek, #2198, BG, 1936, p

YAMHILL COUNTY--6 Bridges

Chehalem Creek, #1998, BG, 1933, p
North Yamhill River, #441, T, 1921, f
North Yamhill (Oak Ridge Road), #6610C, T, 1934, f
Mill Creek, #11815, T, 1918, f
South Yamhill River, #1593, T, 1932, f
South Yamhill River, #2557, T, 1939, f

SUMMARY OF THE INVENTORIED BRIDGES

County	f*	p**	Total	County	f*	p**	Total
Baker	13	—	13	Lake	—	5	5
Benton	14	—	14	Lane	50	12	62
Clackamas	23	—	23	Lincoln	13	1	14
Clatsop	14	5	19	Linn	18	9	27
Columbia	22	16	38	Malheur	12	10	22
Coos	9	1	10	Marion	29	2	31
Crook	3	3	6	Morrow	15	4	19
Curry	5	2	7	Multnomah	64	10	74
Deschutes	2	2	4	Polk	9	3	12
Douglas	29	6	35	Sherman	1	2	3
Gilliam	1	—	1	Tillamook	26	6	32
Grant	1	3	4	Umatilla	36	6	42
Harney	—	—	—	Union	11	4	15
Hood River	12	—	12	Wallowa	8	—	8
Jackson	20	4	24	Wasco	16	5	21
Jefferson	2	2	4	Washington	3	5	8
Josephine	11	5	16	Wheeler	—	1	1
Klamath	5	7	12	Yamhill	5	1	6
				TOTAL	502	142	644

*Field inspected
**Plans inventory only

INDEX

Cape Creek Bridge (1932), Oregon Coast Highway, Heceta Head, Lane County

INDEX

The bridges shown in the photo-description pages of this document are arranged alphabetically by name in this index. Since bridges frequently are identified by several names, all names are included and cross-referenced. Other lists in this document may also be useful in finding a specific bridge. See Appendix C for the bridges listed by location and Appendix D, by structural type.

A

Abiqua Creek (Gallon House) Covered Bridge, Marion County, 166.
Adkisson Bridge--See Fifteenmile Creek (Adkisson) Bridge.
Albany Bridge--See Willamette River (Albany) Bridge.
Alexandra Avenue, N.W., Viaduct, Multnomah County, 94.
Alsea Bay (Waldport) Bridge, Lincoln County, 213.
Alsea River (Hayden) Covered Bridge, Benton County, 168.
Antelope Creek Covered Bridge, Jackson County, 175.
Applegate River (McKee) Covered Bridge, Jackson County, 167.
Astoria Bridge--See Columbia River (Astoria) Bridge.
Auto Club Bridge--See Sandy River (Stark Street) Bridge.

B

Battle Creek Covered Bridge--See Coyote Creek (Battle Creek) Covered Bridge.
Balch Gulch Bridge, Multnomah County, 61.
Beaver Creek (Sandy River Overflow) Bridge, Multnomah County, 122
Beaver Creek (Kukkla Road) Bridge, Columbia County, 275.
Belknap Covered Bridge--See McKenzie River (Belknap) Covered Bridge.
Beltline Overcrossing, Clatsop County, 292.
Ben Jones Bridge--See Rocky Creek (Ben Jones) Bridge.
Big Creek Bridge, Lane County, 283.

Billy Chinook Lake Bridge--See Crooked River (Lake Billy Chinook) Bridge.
Bohemian Hall Covered Bridge--See Crabtree Creek (Bohemian Hall) Covered Bridge.
Booth, Robert A., Bridge--See North Umpqua River (Winchester) Bridge.
Boyd Bridge--See Fifteenmile Creek (Adkisson) Bridge.
Bridal Veil Falls Bridge, Multnomah County, 140.
Bridge of the Gods--See Columbia River (Bridge of the Gods).
Broadway Bridge--See Willamette River (Broadway) Bridge.
Bull Creek Dam Bridge--See Tumalo Irrigation Ditch Bridge.
Bull Run River Bridge, Clackamas County, 59.
Burnside Bridge--See Willamette River (Burnside) Bridge.

C

Calapooia River (Crawfordsville) Covered Bridge, Linn County, 186.
Calapooya Creek (Oakland) Bridge, Douglas County, 277.
Calapooya Creek (Rochester) Covered Bridge, Douglas County, 81.
Cape Creek Bridge, Lane County, 108.
Cascadia Park Bridge--See Santiam River (Cascadia Park) Bridge.
Caveman Bridge--See Rogue River (Caveman) Bridge.
Cavitt Creek Covered Bridge--See Little River (Cavitt Creek) Covered Bridge.
Cecil Bridge--See Willow Creek (Cecil) Bridge.
Chasm (Neahkahnie Mountain) Bridge, Tillamook County, 130.
Chenoweth Creek Bridge, Wasco County, 158.
Chitwood Covered Bridge--See Yaquina River (Chitwood) Covered Bridge.
Church Street, S.E., Bridge--See Pringle Creek/ Shelton Creek (Church Street S.E.) Bridge.
Clackamas River (Estacada) Bridge, Clackamas County, 285.
Clackamas River (McLoughlin) Bridge, Clackamas County, 109.

Clackamas River (Park Place) Bridge, Clackamas County, 276.
Columbia River (Astoria) Bridge, Clatsop County (Oregon) and Pacific County (Washington), 299.
Columbia River (Bridge of the Gods), Hood River County (Oregon) and Skamania County (Washington State), 77.
Columbia River (Interstate Northbound) Bridge, Multnomah County (Oregon) and Clark County (Washington), 209.
Columbia River (Lewis and Clark) Bridge, Columbia County (Oregon) and Cowlitz County (Washington), 212.
Columbia River (White Salmon) Bridge, Hood River County (Oregon) and Klickitat County (Washington State), 288.
Columbia Slough (N.E. Union Avenue) Bridge, Multnomah County, 291.
Commercial Street, South, Bridge--See Pringle Creek (Commercial Street South) Bridge.
Coos Bay (McCullough Memorial) Bridge, Coos County, 82.
Coquille River Bridge, Coos County, 117.
Cow Creek Bridge, Malheur County, 68.
Coyote Creek (Battle Creek) Covered Bridge, Lane County, 176.
Crabtree Creek (Bohemian Hall) Covered Bridge, Linn County, 298.
Crabtree Creek (Hoffman) Covered Bridge, Linn County, 83.
Crabtree Creek (Larwood) Covered Bridge, Linn County, 197.
Crawfordsville Covered Bridge--See Calapooia River (Crawfordsville) Covered Bridge.
Crooked River (Elliott Lane) Bridge, Crook County, 274.
Crooked River (High) Bridge, Jefferson County, 100.
Crooked River (Lake Billy Chinook) Bridge, Jefferson County, 299.
Crow Lane Bridge--See Pine Creek (Crow Lane) Bridge.
Crown Point Viaduct, Multnomah County, 137.
Cummins Creek Bridge, Lane County, 283.
Currin Covered Bridge--See Row River (Currin) Covered Bridge.

D

Dallas Pumping Station Covered Bridge--See Rickreall Creek (Pumping Station) Covered Bridge.
Deadwood Creek Covered Bridge, Lane County, 187.
Depoe Bay Bridge, Lincoln County, 101.
Deschutes River (Maupin) Bridge, Wasco County, 278.
Devaney Covered Bridge--See Thomas Creek (Weddle) Covered Bridge.
Dollarhide Overcrossing, Jackson County, 123.
Dorena Covered Bridge--See Row River (Dorena) Covered Bridge.
Drift Creek Covered Bridge, Lincoln County, 164.
Dry Canyon Creek Bridge, Wasco County, 159.

E

Eagle Creek Bridge, Multnomah County, 150.
Eagle Creek and Toothrock viaducts--See Toothrock and Eagle Creek viaducts.
Earnest Covered Bridge--See Mohawk River (Earnest) Covered Bridge.
East Multnomah Falls Viaduct, Multnomah County, 144.
Eighth Street, S.E., Bridge--See Umatilla River (S.E. 8th Street) Bridge.
Elk Creek (First Crossing) Bridge, Douglas County, 279.
Elk Creek (Second Crossing) Bridge, Douglas County, 294.
Elk Creek (Third Crossing) Bridge, Douglas County, 280.
Elk Creek (Fourth Crossing) Bridge, Douglas County, 280.
Elk Creek (Roaring Camp) Covered Bridge, Douglas County, 183.
Elliott Lane Bridge--See Crooked River (Elliott Lane) Bridge.
Estacada Bridge--See Clackamas River (Estacada) Bridge.
Euchre Creek Bridge, Curry County, 293.
Evans Creek (Wilmer) Covered Bridge, Jackson County, 180.

F

Fall Creek (Pengra) Covered Bridge, Lane County, 192.
Fall Creek (Unity) Covered Bridge, Lane County, 188.
Fessenden, North, Street Overcrossing--See North Fessenden Street Overcrossing.
Fifteenmile Creek (Adkisson) Bridge, Wasco County, 99.
Fifteenmile Creek (Seufert) Viaduct, Wasco County, 126.
First Crossing Bridge--See Elk Creek (First Crossing) Bridge.
Fisher School Covered Bridge--See Five Rivers (Fisher School) Covered Bridge.
Five Rivers (Fisher School) Covered Bridge, Lincoln County, 169.
Florence Bridge--See Siuslaw River (Florence) Bridge.
Fourth Crossing Bridge--See Elk River (Fourth Crossing) Bridge.
Fremont Bridge--See Willamette River (Fremont) Bridge.
Front Street, N.E., Bridge--See Mill Creek (Front Street N.E.) Bridge.

G

Gallon House Covered Bridge--See Abiqua Creek (Gallon House) Covered Bridge.
Gilkey Covered Bridge--See Thomas Creek (Gilkey) Covered Bridge.
Gold Beach Bridge--See Rogue River (Gold Beach) Bridge.
Gold Hill Bridge--See Rogue River (Gold Hill) Bridge.
Goodpasture Covered Bridge--See McKenzie River (Goodpasture) Covered Bridge.
Gorton Creek Bridge, Hood River County, 152.
Grande Avenue, N.E., Overcrossing, Multnomah County, 289.
Grande Ronde River (McKennon Road) Bridge, Union County, 278.
Grande Ronde River (Old Rhinehart) Bridge, Union County, 74.
Grande Ronde River (Perry Overcrossing) Bridge, Union County, 281.
Grande Ronde River (Troy) Bridge, Wallowa County, 69.
Grande Ronde River (Yarrington) Bridge, Union County, 269.
Grant Road Bridge--See Luckiamute River (Grant Road) Bridge.
Grave Creek Covered Bridge, Josephine County, 171.
Greasewood Creek Bridge, Umatilla County, 269.

H

Hannah Covered Bridge--See Thomas Creek (Hannah) Covered Bridge.
Harris Covered Bridge--See Marys River (Harris) Covered Bridge.
Hawthorne Bridge--See Willamette River (Hawthorne) Bridge.
Hayden Covered Bridge--See Alsea River (Hayden) Covered Bridge.
High Bridge--See Crooked River (High) Bridge.
Hoffman Covered Bridge--See Crabtree Creek (Hoffman) Covered Bridge.
Hog Creek Canyon (Rowena Dell) Bridge, Wasco County, 157.
Hood River, South Fork, Bridge--See South Fork Hood River (Sahalie Falls) Bridge.
Hood River (Tucker) Bridge, Hood River County, 284.
Horse Creek Covered Bridge, Lane County, 184.
Horsetail Falls Bridge, Multnomah County, 146.

I

Interstate Bridge--See Columbia River (Interstate Northbound) Bridge.
Irish Bend Covered Bridge--See Willamette Slough (Irish Bend) Covered Bridge.

J

Jefferson Bridge--See Santiam River (Jefferson) Bridge.
Johnson Creek Bridge, Multnomah County, 290.
Jones, Ben, Bridge--See Rocky Creek (Ben Jones) Bridge.
Jordan Covered Bridge--See Thomas Creek (Jordan) Covered Bridge.

Jordan Creek Overflow Bridge, Malheur County, 271.

K

Killam Creek Bridge, Tillamook County, 89.
Kukkla Road Bridge--See Beaver Creek (Kukkla Road) Bridge.

L

Lake Billy Chinook Bridge--See Crooked River (Lake Billy Chinook) Bridge.
Lake Creek (Nelson Mountain) Covered Bridge, Lane County, 182.
Larwood Covered Bridge--See Crabtree Creek (Larwood) Covered Bridge.
Latourell Creek Bridge, Multnomah County, 138.
Lewis and Clark Bridge--See Columbia River (Lewis and Clark) Bridge.
Lewis and Clark River Bridge, Clatsop County, 288.
Liberty Street, Bridge--See Pringle Creek (Liberty Street S.E.) Bridge.
Link River Bridge, Klamath County, 295.
Little River (Cavitt Creek) Covered Bridge, Douglas County, 297.
Lombard, North, Street Overcrossing--See North Lombard Street Overcrossing.
Longview Bridge--See Columbia River (Lewis and Clark) Bridge
Lost Creek Covered Bridge, Jackson County, 170.
Lost Creek (Parvin) Covered Bridge, Lane County, 173.
Lowell Covered Bridge--See Middle Fork Willamette River (Lowell) Covered Bridge.
Lower Milton Creek (McDonald) Bridge, Columbia County, 73.
Luckiamute River (Grant Road) Bridge, Polk County, 275.
Lusted Road Bridge--See Sandy River (Lusted Road) Bridge.

M

Marys River Bridge, Benton County, 274.
Marys River (Harris) Covered Bridge, Benton County, 189.
Maupin Bridge--See Deschutes River (Maupin) Bridge.

McCarthy Creek Bridge, Multnomah County, 290.
McCullough Memorial Bridge--See Coos Bay (McCullough Memorial) Bridge.
McDonald Bridge--See Lower Milton Creek (McDonald) Bridge.
McKee Covered Bridge--See Applegate River (McKee) Covered Bridge.
McKennon Road Bridge--See Grande Ronde River (McKennon Road) Bridge.
McKenzie River (Belknap) Covered Bridge, Lane County, 205.
McKenzie River (Goodpasture) Covered Bridge, Lane County, 193.
McLoughlin Bridge--See Clackamas River (McLoughlin) Bridge.
Middle Fork Willamette River (Lowell) Covered Bridge, Lane County, 199.
Mill Creek (West Sixth Street) Bridge, Wasco County, 127.
Mill Creek (Front Street N.E.) Bridge, Marion County, 88.
Mill Creek (Summer Street N.E.) Bridge, Marion County, 294.
Mill Creek (Wendling) Covered Bridge, Lane County, 194.
Mill, Old, Race Bridge--See Old Mill Race Bridge.
Milo Academy Covered Bridge--See South Umpqua River (Milo Academy) Covered Bridge.
Moffett Creek Bridge, Multnomah County, 147.
Mohawk River (Ernest) Covered Bridge, Lane County, 195.
Mosby Creek Covered Bridge, Lane County, 172.
Mosby Creek (Stewart) Covered Bridge, Lane County, 185.
Mosier Creek Bridge, Wasco County, 156.
Mosier Creek (State Road) Bridge, Wasco County, 292.
Mott Bridge--See North Umpqua River (Mott) Bridge.
Multnomah Falls Bridge--See Multnomah Creek Bridge.
Multnomah Creek Bridge, Multnomah County, 143.
Multnomah Falls, East, Viaduct--See East Multnomah Falls Viaduct.
Multnomah Falls, West, Viaduct--See West Multnomah Falls Viaduct.

Myrtle Creek Bridge--See South Umpqua River (Myrtle Creek) Bridge.
Myrtle Creek, (Neal Lane) Covered Bridge--See South Myrtle Creek (Neal Lane) Covered Bridge.
Myrtle Street Bridge--See Powder River (Myrtle Street) Bridge.

N

Neahkahnie Mountain Bridge--See Chasm (Neahkahnie Mountain) Bridge.
Neal Lane Covered Bridge--See South Myrtle Creek (Neal Lane) Covered Bridge.
Nelson Covered Bridge--See Lake Creek (Nelson Mountain) Covered Bridge.
Nelson Mountain Covered Bridge--See Lake Creek (Nelson Mountain) Covered Bridge.
Necanicum River (Seaside) Bridge, Clatsop County, 97.
Necarney Creek Bridge, Tillamook County, 131.
Nehalem River Bridge, Clatsop County, 286.
Newport Bridge--See Yaquina Bay (Newport) Bridge.
North Bend Bridge--See Coos Bay (McCullough Memorial) Bridge.
North Fessenden Street Overcrossing, Multnomah County, 63.
North Fork of the Middle Fork Willamette River (Office) Covered Bridge, Lane County, 198.
North Fork Yachats River Covered Bridge, Lincoln County, 196.
North Lombard Street Overcrossing, Multnomah County, 270.
North Umpqua River (Mott) Bridge, Douglas County, 214.
North Umpqua River (Winchester) Bridge, Douglas County, 98.
North Willamette Boulevard Overcrossing, Multnomah County, 64.
North Yamhill River Bridge, Yamhill County, 277.
Northeast 12th Avenue Overcrossing, Multnomah County, 289.

O

Oakland Bridge--See Calapooya Creek (Oakland) Bridge.
Office Covered Bridge--See North Fork of the Middle Fork Willamette River (Office) Covered Bridge.
Old Mill Race Bridge, Umatilla County, 124.
Old Rhinehart Bridge--See Grande Ronde River (Old Rhinehart) Bridge.
Old Young's Bay Bridge, Clatsop County, 287.
Olex Bridge--See Rock Creek (Olex) Bridge.
Oneonta Gorge Creek Old Bridge, Multnomah County, 145.
Oneonta Gorge Creek New Bridge, Multnomah County, 160.
Oregon City-West Linn Bridge--See Willamette River (Oregon City) Bridge.
Oswego Creek Bridge, Clackamas County, 210.
Owyhee River Bridge, Malheur County, 62.

P

Park Place Bridge--See Clackamas River (Park Place) Bridge.
Parvin Covered Bridge--See Lost Creek (Parvin) Covered Bridge.
Pass Creek Covered Bridge, Douglas County, 75.
Pengra Covered Bridge--See Fall Creek (Pengra) Covered Bridge.
Perry Overcrossing--See Grande Ronde River (Perry Overcrossing) Bridge.
Pine Creek Bridge, (No. 10712), Umatilla County (near Weston), 272.
Pine Creek Bridge, (No. 59C528), Umatilla County (near Umapine), 271.
Pine Creek Bridge, (No. 59C534), Umatilla County (near Umapine), 70.
Pine Creek (Crow Lane) Bridge, Baker County, 273.
Portland Road, N.E., Undercrossing, Marion County, 295.
Powder River (Myrtle Street) Bridge, Baker County, 276.
Pringle Creek (Commercial Street South) Bridge, Marion County, 293.
Pringle Creek (Liberty Street S.E.) Bridge, Marion County, 128.
Pringle Creek/Shelton Creek (Church Street S.E.) Bridge, Marion County, 129.
Pumping Station Covered Bridge--See Rickreall Creek (Pumping Station) Covered Bridge.

R

Reed, Samuel G., Bridge--See Necarney Creek Bridge.
Reedsport Bridge--See Umpqua River (Reedsport) Bridge.
Remote Covered Bridge--See Sandy Creek (Remote) Covered Bridge.
Rhea Creek Bridge, Morrow County (near Jordan), 291.
Rhea Creek (Spring Hollw) Bridge, Morrow County, 65.
Richardson Bridge--See Siuslaw River (Richardson) Bridge.
Richardson Gap Road Covered Bridge--See Crabtree Creek (Bohemian Hall) Covered Bridge.
Rickreall Creek (Pumping Station) Covered Bridge, Polk County, 165.
Ritner Creek Covered Bridge, Polk County, 181.
Roaring Camp Covered Bridge--See Elk Creek (Roaring Camp) Covered Bridge.
Robertson Bridge--See Rogue River (Robertson) Bridge.
Rochester Covered Bridge--See Calapooya Creek (Rochester) Covered Bridge.
Rock Creek Bridge, Wasco County, 154.
Rock Creek (Olex) Bridge, Gilliam County, 207.
Rock O' the Range Covered Bridge--See Swalley Canal (Rock O' the Range) Covered Bridge.
Rock Point Bridge--See Rogue River (Rock Point) Bridge.
Rock Slide Viaduct, Hood River County, 155.
Rocky Creek (Ben Jones) Bridge, Lincoln County, 102.
Rogue River (Caveman) Bridge, Josephine County, 105.
Rogue River (Gold Beach) Bridge, Curry County, 106.
Rogue River (Gold Hill) Bridge, Jackson County, 103.
Rogue River (Robertson) Bridge, Josephine County, 270.
Rogue River (Rock Point) Bridge, Jackson County, 93.
Ross Island Bridge--See Willamette River (Ross Island) Bridge.
Row River (Currin) Covered Bridge, Lane County, 177.
Row River (Dorena) Covered Bridge, Lane County, 201.
Rowena Dell Bridge--See Hog Creek Canyon (Rowena Dell) Bridge.
Ruckel Creek Bridge, Hood River County, 151.
Ruthton Point Viaduct, Hood River County, 153.

S

Sahalie Falls Bridge--See South Fork Hood River (Sahalie Falls) Bridge.
Salmon River Bridge, Lincoln County, 282.
Samuel G. Reed Bridge--See Necarney Creek Bridge.
Sandy River Overflow Bridge--See Beaver Creek (Sandy River Overflow) Bridge.
Sandy River (Lusted Road) Bridge, Clackamas County, 60.
Sandy River (Remote) Covered Bridge, Coos County, 174.
Sandy River (Stark Street) Bridge, Multnomah County, 136.
Sandy River (Troutdale) Bridge, Multnomah County. 135.
Santiam River bridges (I-5), Marion-Linn counties, 297
Santiam River (Cascadia Park) Bridge, Linn County, 79.
Santiam River (Jefferson) Bridge, Marion-Linn counties, 110.
Santiam River, South Fork, (Short) Covered Bridge--See South Fork Santiam River (Short) Covered Bridge.
Scottsburg Bridge--See Umpqua River (Scottsburg) Bridge.
Second Crossing Bridge--See Elk River (Second Crossing) Bridge.
Seaside Bridge--See Necanicum River (Seaside) Bridge.
Seufert Viaduct--See Fifteenmile Creek (Seufert) Viaduct.
Shelton Creek/Pringle Creek Bridge--See Pringle Creek/Shelton Creek (Church Street S.E.) Bridge.

Shepperd's Dell Bridge--See Young Creek (Shepperd's Dell) Bridge.
Shimanek Covered Bridge--See Thomas Creek (Shimanek) Covered Bridge.
Short Covered Bridge--See South Fork Santiam River (Short) Covered Bridge.
Siuslaw River (Florence) Bridge, Lane County, 119.
Siuslaw River (Richardson) Bridge, Lane County, 71.
Sixth Street, West, Bridge--See Mill Creek (West Sixth Street) Bridge.
Soapstone Creek Bridge, Clatsop County, 282.
Southern Pacific Railroad (N.E. Portland Road) Overcrossing--See Portland Road, N.E., Undercrossing.
South Commercial Street Bridge--See Pringle Creek (Commercial Street S.E.) Bridge.
South Fork Hood River (Sahalie Falls) Bridge, Hood River County, 104.
South Fork Santiam River (Short) Covered Bridge, Linn County, 200.
South Myrtle Creek (Neal Lane) Covered Bridge, Douglas County, 84.
South Umpqua River (Milo Academy) Covered Bridge, Douglas County, 203.
South Umpqua River (Myrtle Creek) Bridge, Douglas County, 95.
South Umpqua River (Winston) Bridge, Douglas County, 285.
South Umpqua River (Worthington) Bridge, Douglas County, 272.
South Yamhill River (Steel) Bridge, Polk County, 273.
Spokane, Portland and Seattle Railroad Undercrossing (North Fessenden Street)--See North Fessenden Street Overcrossing.
Spokane, Portland and Seattle Railroad Undercrossing (North Lombard Street)--See North Lombard Street Overcrossing.
Spokane, Portland and Seattle Railroad Undercrossing (North Willamette Street)--See North Willamette Street Overcrossing.
Springfield Bridge--See Willamette River (Springfield) Bridge.
Spring Hollow Bridge--See Rhea Creek (Spring Hollow) Bridge.

St. John's Bridge--See Willamette River (St. John's) Bridge.
Star Covered Bridge--See Row River (Dorena) Covered Bridge.
Stark Street Bridge--See Sandy River (Stark Street) Bridge.
Stark Street Viaduct, Multnomah County, 91.
State Road Bridge--See Mosier Creek (State Road) Bridge.
Steel Bridge--See South Yamhill River (Steel) Bridge.
Steel Bridge--See Willamette River (Steel) Bridge.
Steinman Overcrossing, Jackson County, 125.
Stewart Covered Bridge--See Mosby Creek (Stewart) Covered Bridge.
Sucker Creek Bridge--See Oswego Creek Bridge.
Summer, N.E., Street Bridge--See Mill Creek (N.E. Summer Street) Bridge.
S.W. Vista Ave. Viaduct, Multnomah Co., 211.
Swalley Canal (Rock O' the Range) Covered Bridge, Deschutes County, 204.
Swing Log Covered Bridge--See Coyote Creek (Battle Creek) Covered Bridge.

T

Tanner Creek Bridge, Multnomah County, 148.
Tenmile Creek Bridge, Lane County, 284.
Tenth Street, S.W., Bridge--See Umatilla River (S.W. 10th Street) Bridge.
Third Crossing Bridge--See Elk Creek (Third Crossing) Bridge.
Thomas Creek Bridge, Curry County, 298.
Thomas Creek (Gilkey) Covered Bridge, Linn County, 85.
Thomas Creek (Hannah) Covered Bridge, Linn County, 190.
Thomas Creek (Jordan) Covered Bridge, Linn County, 215.
Thomas Creek (Shimanek) Covered Bridge, Linn County, 300.
Thomas Creek (Weddle) Covered Bridge, Linn County, 191.
Toothrock and Eagle Creek viaducts, Multnomah County, 149.
Troutdale Bridge--See Sandy River (Troutdale) Bridge.

Troy Bridge--See Grande Ronde River (Troy) Bridge.
Tucker Bridge--See Hood River (Tucker) Bridge.
Tumalo Irrigation Ditch Bridge, Deschutes County, 90.

U

Umatilla Bridge--See Umatilla River (Umatilla) Bridge.
Umatilla River (S.E. 8th Street) Bridge, Umatilla County, 66.
Umatilla River (S.W. 10th Street) Bridge, Umatilla County, 72.
Umatilla River (Umatilla) Bridge, Umatilla County, 281.
Umpqua River, North, (Mott) Bridge--See North Umpqua River (Mott) Bridge.
Umpqua River, North, (Winchester) Bridge--See North Umpqua River (Winchester) Bridge.
Umpqua River (Reedsport) Bridge, Douglas County, 120.
Umpqua River (Scottsburg) Bridge, Douglas County, 279.
Umpqua River (Milo Academy) Covered Bridge -- See South Umpqua River (Milo Academy) Covered Bridge.
Umpqua River (Myrtle Creek) Bridge--See South Umpqua River (Myrtle Creek) Bridge.
Umpqua River (Winston) Bridge--See South Umpqua River (Winston) Bridge.
Umpqua River (Worthington) Bridge--See South Umpqua River (Worthington) Bridge.
Union Avenue, N.E., Bridge--See Columbia Slough (N.E. Union Avenue) Bridge.
Unity Covered Bridge--See Fall Creek (Unity) Covered Bridge.

V

Van Buren Street Bridge--See Willamette River (Van Buren Street) Bridge.
Vista Avenue Viaduct--See S.W. Vista Avenue Viaduct.

W

Wahkeena Falls Bridge, Multnomah County, 141.
Waldport Bridge--See Alsea Bay (Waldport) Bridge.
Weddle Covered Bridge--See Thomas Creek (Weddle) Covered Bridge.
Wendling Covered Bridge--See Mill Creek (Wendling) Covered Bridge.
West Broadway Street Bridge--See Necanicum River (Seaside) Bridge.
West Linn-Oregon City Bridge--See Willamette River (Oregon City) Bridge.
West Multnomah Falls Viaduct, Multnomah County, 142.
West Sixth Street Bridge--See Mill Creek (West Sixth Street) Bridge.
Whiskey Butte Covered Bridge--See South Fork Santiam River (Short) Covered Bridge.
White Salmon Bridge--See Columbia River (White Salmon) Bridge.
Wildcat Creek Covered Bridge, Lane County, 178.
Willamette, North, Boulevard Overcrossing--See North Willamette Boulevard Overcrossing.
Willamette River (Albany) Bridge, Linn-Benton counties, 76.
Willamette River (Broadway) Bridge, Multnomah County, 116.
Willamette River (Burnside) Bridge, Multnomah County, 118.
Willamette River (Fremont) Bridge, Multnomah County, 300.
Willamette River (Hawthorne) Bridge, Multnomah County, 115.
Willamette River, Middle Fork, (Lowell) Covered Bridge--See Middle Fork Willamette River (Lowell) Covered Bridge.
Willamette River, North Fork of the Middle Fork, (Office) Covered Bridge--See North Fork of the Middle Fork Willamette River (Office) Covered Bridge.
Willamette River (Oregon City) Bridge, Clackamas County, 96.

Willamette River (Ross Island) Bridge, Multnomah County, 78.
Willamette River (Springfield) Bridge, Lane County, 80.
Willamette River (Steel) Bridge, Multnomah County, 208.
Willamette River (St. John's) Bridge, Multnomah County, 113.
Willamette River (Van Buren Street) Bridge, Benton-Linn counties, 287.
Willamette Slough (Irish Bend) Covered Bridge, Benton County, 202.
Williams Creek Bridge, Josephine County, 92.
Willow Creek (Cecil) Bridge, Morrow County, 67.
Wilson River Bridge, Tillamook County, 107.
Wimer Covered Bridge--See Evans Creek (Wimer) Covered Bridge.
Winchester Bridge--See North Umpqua River (Winchester) Bridge.
Winston Bridge--See South Umpqua River (Winston) Bridge.
Worthington Bridge--See South Umpqua River (Worthington) Bridge.

Y

Yachats River, North Fork, Covered Bridge--See North Fork Yachats River, Covered Bridge.
Yamhill, North, River Bridge--See North Yamhill River Bridge.
Yamhill, South, River (Steel) Bridge--See South Yamhill River (Steel) Bridge.
Yaquina Bay (Newport) Bridge, Lincoln County, 111.
Yaquina River (Chitwood) Covered Bridge, Lincoln County, 179.
Yarrington Bridge--See Grande Ronde River (Yarrington) Bridge.
Young Creek (Shepperd's Dell) Bridge, Multnomah County, 139.
Young's Bay, Old, Bridge--See Old Young's Bay Bridge.

PHOTOGRAPHY CREDITS

Josephine County Historical Society	Fig. 7.
Jef Kaiser	p. 209.
Roger Keiffer	Fig. 14.
Craig Markham	Fig. 2b; pp. 61, 91, 94, 102, 107, 109, 212, 276, 276b, 277a, 282, 284a, 285, 286, 287b, 288b, 299b.
James Norman	Fig. 2c, Fig. 10, Fig. 16, Fig. 18a, Fig. 19, Fig. 22, Fig. 25; pp. 59, 60, 63, 64, 65, 66, 67, 70, 71, 72, 73, 74, 75, 76, 77, 78, 80, 81, 83, 84, 85, 88, 90, 92, 93, 96, 98, 103, 104, 105, 108, 110, 113, 115, 116, 117, 118, 122, 123, 124, 125, 128, 129, 130, 131, 137, 138, 139, 140, 141, 142, 143, 144, 145, 146, 148, 150, 151, 152, 153, 154, 155, 157, 160, 164, 165, 166, 167, 168, 169, 170, 171, 172, 173, 174, 175, 176, 177, 178, 179, 180, 181, 182, 183, 184, 185, 186, 187, 188, 189, 190, 191, 192, 193, 194, 195, 196, 197, 198, 199, 200, 201, 202, 203, 204, 205, 207, 208, 214, 215, 221, 269, 270, 271, 282, 273a, 274, 275b, 278a, 279, 280, 281a, 287a, 289, 290, 291, 292, 293, 294, 295, 297, 298, 299a, 300a.
ODOT Bridge Section Collection	Fig. 5, Fig. 6, Fig. 9, Fig. 11, Fig. 13, Fig. 17, Fig. 24, p. 220.
ODOT Photography Lab Files	Fig. 1, Fig. 12; pp. 111, 120.
John Preston	p. 281b.
Jerry Robertson	Fig. 2a, Fig. 4, Fig. 18; pp. 62, 68, 82, 100, 101, 106, 119, 210, 213, 300b.
Dwight Smith	Fig. 18d; pp. 69, 79, 89, 97, 99, 126, 127, 135, 136, 147, 156, 158, 159, 273b, 275a, 276a, 277b, 278b, 284b, 288a.
Murray Stone	p. 211.

The photographs in this document are the property of the Oregon Department of Transportation with one exception. The early Rogue River Bridge photo (Figure 7) was furnished courtesy of the Josephine County Historical Society.

NOTE ON THE DIVIDERS

The major sections of this document are divided by drawings of early bridge railing designs in Oregon from the turn of the century to the late 1930s. In addition to being functional, bridge railings provide an opportunity for adding artistic treatment to structures and contribute to the aesthetic appeal of structures. In some cases, the railing is the only feature of a bridge readily seen by the traveler and often only in a brief glimpse during the crossing. Railing designs have changed over time, in conjunction with changes in bridge technology, materials, styles, taste, and safety. Some designs were created specifically for one bridge, while others were standardized and used on hundreds of bridges. (Drawings by George Kraus, ODOT.)

COLOPHON

As part of the Oregon Historical Society Press commitment to keep important previously published historical studies in print, *Historic Highway Bridges of Oregon* has been published in this paperbound edition.

A new cover and title page were designed by the OHS Press for this edition, and slight changes to the look of the interior pages have been made. Subsequent to the publication of the first edition in 1985, some significant developments affecting specific bridges around the state have required changes in the text. Except for those changes, the content of this edition is a full and true facsimile of the 1985 version, which was written and published by the Oregon Department of Transportation.

Souvenir, a typeface common to many books issued over the past two decades, was used for both the text and display typography. It was set by the State of Oregon Printing Division. Printing, on a seventy-pound coated matte offset paper, and binding of this edition were done in Ann Arbor, Michigan, by Malloy Lithographing, Inc.